SpringerBriefs in Geoethics

Editor-in-Chief

Silvia Peppoloni, National Institute of Geophysics and Volcanology (INVG), Rome, Italy

Series Editors

Nic Bilham, University of Exeter, Penryn, UK

Peter T. Bobrowsky, Geological Survey of Canada, Sidney, Canada

Vincent S. Cronin, Baylor University, Waco, USA

Giuseppe Di Capua, National Institute of Geophysics and Volcanology (INVG), Rome, Italy

Iain Stewart, University of Plymouth, Plymouth, UK

Artur Sá, University of Trás-os-Montes and Alto Douro, Vila Real, Portugal

Rika Preiser, Stellenbosch University, Stellenbosch, South Africa

SpringerBriefs in Geoethics envisions a series of short publications that aim to discuss ethical, social, and cultural implications of geosciences knowledge, education, research, practice and communication. The series SpringerBriefs in Geoethics is sponsored by the IAPG – International Association for Promoting Geoethics (http://www.geoethics.org).

The intention is to present concise summaries of cutting-edge theoretical aspects, research, practical applications, as well as case-studies across a wide spectrum.

SpringerBriefs in Geoethics are seen as complementing monographs and journal articles, or developing innovative perspectives with compact volumes of 50 to 125 pages, covering a wide range of contents comprising philosophy of geosciences and history of geosciences thinking; research integrity and professionalism in geosciences; working climate issues and related aspects; geoethics in georisks and disaster risk reduction; responsible georesources management; ethical and social aspects in geoeducation and geosciences communication; geoethics applied to different geoscience fields including economic geology, paleontology, forensic geology and medical geology; ethical and societal relevance of geoheritage and geodiversity; sociological aspects in geosciences and geosciences-society-policy interface; geosciences for sustainable and responsible development; geoethical implications in global and local changes of socio-ecological systems; ethics in geoengineering; ethical issues in climate change and ocean science studies; ethical implications in geosciences data life cycle and big data; ethical and social matters in the international geoscience cooperation.

Typical topics might include:

– Presentations of core concepts
– Timely reports on state-of-the art
– Bridges between new research results and contextual literature reviews
– Innovative and original perspectives
– Snapshots of hot or emerging topics
– In-depth case studies or examples

All projects will be submitted to the series-editor for consideration and editorial review.

Each volume is copyrighted in the name of the authors. The authors and IAPG retain the right to post a pre-publication version on their respective websites.

The Series in Geoethics is initiated and supervised by Silvia Peppoloni and an editorial board formed by Nic Bilham, Peter T. Bobrowsky, Vincent S. Cronin, Giuseppe Di Capua, Rika Preiser, Artur Agostinho de Abreu e Sá, Iain Stewart.

Silvia Peppoloni · Giuseppe Di Capua

Geoethics and Geosciences Serving Society

Reflections from the 37th International Geological Congress

 Springer

Silvia Peppoloni (ORCID)
Istituto Nazionale di Geofisica e
Vulcanologia
Rome, Italy

International Association for Promoting
Geoethics (IAPG)
Rome, Italy

Chair on Geoethics of the International
Council for Philosophy and Human
Sciences (CIPSH-CG)
Rome, Italy

Commission on Geoethics
of the International Union of Geological
Sciences (IUGS-CG)
Rome, Italy

Giuseppe Di Capua (ORCID)
Istituto Nazionale di Geofisica e
Vulcanologia
Rome, Italy

International Association for Promoting
Geoethics (IAPG)
Rome, Italy

Chair on Geoethics of the International
Council for Philosophy and Human
Sciences (CIPSH-CG)
Rome, Italy

Commission on Geoethics
of the International Union of Geological
Sciences (IUGS-CG)
Rome, Italy

ISSN 2662-6780   ISSN 2662-6799   (electronic)
SpringerBriefs in Geoethics
ISBN 978-3-032-03753-4   ISBN 978-3-032-03754-1   (eBook)
https://doi.org/10.1007/978-3-032-03754-1

This Springer imprint is published by the registered company Springer Nature Switzerland AG
The registered company address is: Gewerbestrasse 11, 6330 Cham, Switzerland

If disposing of this product, please recycle the paper.

Contents

Introduction

Silvia Peppoloni and **Giuseppe Di Capua**

Abstract This book brings together reflections from the session "Geoethics at the Heart of All Geoscience: Serving the Public Good," held under the theme "Geoethics and Societal Relevance of Geosciences" at the 37th International Geological Congress (IGC) in Busan, Republic of Korea. The session was co-sponsored by the International Association for Promoting Geoethics (IAPG) and the Commission on Geoethics (CG) of the International Union of Geological Sciences (IUGS). The wealth of contributions presented, rooted not only in the broad field of geosciences but also intersecting with diverse scientific, humanistic, and social disciplines, makes this volume a representative and insightful compendium. It exemplifies the cultural and technical-scientific diversity of perspectives that geoethics inspires within the geoscientific community, highlighting its ability to foster interdisciplinary dialogue and address complex global challenges through an ethical lens.

Keywords Geoethics · Cultural diversity · Interdisciplinarity · Sustainability · Governance

1 The Goal of This Book

This important book continues the ongoing effort to develop a more nuanced and precise framework for geoethics. It offers reflections on geoscientific issues aimed at raising awareness within the geoscientific community, among scholars from other disciplines, and across society at large, about the cultural, scientific, and technical

S. Peppoloni (✉) · G. Di Capua
Istituto Nazionale di Geofisica e Vulcanologia, Rome, Italy
e-mail: silvia.peppoloni@ingv.it

International Association for Promoting Geoethics (IAPG), Rome, Italy

Chair on Geoethics of the International Council for Philosophy and Human Sciences (CIPSH-CG), Rome, Italy

Commission on Geoethics of the International Union of Geological Sciences (IUGS-CG), Rome, Italy

S. Peppoloni and G. Di Capua, *Geoethics and Geosciences Serving Society*, SpringerBriefs in Geoethics, https://doi.org/10.1007/978-3-032-03754-1_1

significance of geoscience research and applications. Additionally, the book provides practical tools for applying the principles and values of geoethics, with the ultimate goal of promoting human well-being and fostering a renewed understanding of the relationship between humanity and the Earth system.

2 An Overview on Chapters

This volume explores geoethics as a comprehensive ethical approach to the geosciences, addressing critical issues such as sustainability, environmental and social justice, the management of uncertainty, and the individual and collective responsibilities of geoscientists. The contributions cover a wide range of topics—from a methodological framework for geoethics and the importance of cultural diversity in preventing geoscientific groupthink, to the integration of Indigenous knowledge and the role of scientific diplomacy. It also addresses ethics in geohazard research, the significance of geotourism and geoconservation, and reflections on the Anthropocene and the ethics of human activities in outer space. The book emphasizes the necessity of an interdisciplinary and culturally sensitive perspective to effectively confront global challenges through conscious and responsible scientific action.

The chapters are arranged to provide a compelling narrative, progressing from foundational concepts of geoethics to diverse practical applications and increasingly broader scales of ethical consideration, ultimately culminating in diplomacy—a forward-looking perspective on how geoethics can be implemented in practice.

The first chapter (Peppoloni et al., 2025) introduces Ecological Humanism as a vision that promotes a balanced relationship between humans and nature. It proposes Relational Geoscientific Pragmatism (RGP) as a methodological framework for geoethical action, emphasizing pragmatic and context-sensitive solutions, interdisciplinary collaboration, and the integration of science, ethics, and social values. The chapter addresses operational challenges in geoethics, such as balancing universal principles with local specificities and managing uncertainty in decision-making. It also outlines urgent priorities for applying geoethics, including the responsible use of technologies and data, the promotion of environmental and social justice, and the advancement of geoethical education.

Butler et al. (2025) examine the vital importance of valuing interpretive diversity in geological studies, using the 1920s Continental Drift debate as a historical example. They argue that aggressive rhetoric and groupthink among some North American geologists marginalized alternative interpretations, delaying progress in global tectonics research for decades. Their work emphasizes that recognizing and respecting diverse perspectives, as well as fostering greater diversity in scientific debates, is essential to prevent similar setbacks and to improve the assessment of uncertainty in geological interpretations.

Further enriching the discussion, Hughes and Kereikeepa (2025) highlight how traditional knowledge from Indigenous communities can profoundly enhance geoethics, focusing on Māori beliefs in Aotearoa New Zealand. Their chapter explores Māori animism and holism, which attribute a spiritual vital essence (*mauri*) and agency to landscape features, underpinning an ethics of guardianship (*kaitiakitanga*) and sustainability. Through a case study of post-earthquake reconstruction in Kaikōura, they illustrate how infrastructure projects can sometimes cause greater harm to sacred Indigenous landscapes (*wāhi tapu*) than the natural disasters themselves. This chapter emphasizes the geoethical responsibility to balance societal needs with the rights and interests of Indigenous communities, highlighting the importance of individual ethical action by geoscientists and engineers, even within restrictive legal frameworks.

Kelman (2025) focuses on applied ethics in geohazard research, navigating the delicate balance between scientific inquiry and the safety of all stakeholders. The chapter highlights various physical, social, and ethical risks that can arise during research and publication, including exposure to natural disasters, dangers faced by citizen scientists in authoritarian regimes, and the impact of risk communication on property values. It also addresses the challenges of managing scientific misinformation in geohazard studies and examines how codes of conduct and science diplomacy can enhance operational ethics, while acknowledging the difficulties involved in their implementation and the need for accountability.

The chapter by Khan et al. (2025) applies geoethical principles to geotourism and geoconservation, focusing on the Sundarbans wetlands in India. It presents an inventory of geosites evaluated using the M-GAM model to assess their geotourism potential for promoting rural development. The authors emphasize that geosite management carries significant geoethical responsibilities, including supporting sustainable development, protecting natural heritage (geodiversity, biodiversity, and cultural diversity), and raising public awareness, all of which are essential for the conservation and well-being of local communities and wildlife.

Koster et al. (2025) discuss the concept of the Anthropocene, analyzing the scientific and ethical controversy surrounding its formal definition as a geological epoch versus its broader understanding as a complex event. They stress the need for a geoethical foundation in Anthropocene discourse, arguing that the concept of the "Anthropocene Event", which adopts a holistic Earth–Human systems perspective, offers a more suitable framework for Earth Governance to address global crises. The authors also critique the unconventional approach of the Anthropocene Working Group (AWG) and discuss the ethical implications of using nuclear weapons as a stratigraphic marker.

In the chapter by Greenbaum (2025), issues concerning human action in an extraterrestrial context are addressed. The chapter introduces exoethics as an ethical framework for human activities beyond Earth, extending the geoethical perspective to extraterrestrial contexts with a specific focus on terraforming Mars. It outlines core exoethical principles, including Planetary Dignity and Intrinsic Value, Precautionary Stewardship, Distributive Cosmic Justice, Life-Centered Ethics, and Governance/Accountability. The chapter further discusses the ethical, legal, and governance challenges associated with terraforming, such as biological contamination, dual-use technology risks, and potential monopolization. Through case studies, it offers ethical evaluations of various terraforming proposals, including local atmospheric modification, solar mirrors, and bioengineered extremophiles.

Finally, the chapter by Raji and Awosika (2025) defines geoscientific diplomacy as the vital bridge between geoscientific expertise and international policy, enabling collaboration and informed decision-making. Their chapter highlights the Commission on the Limits of the Continental Shelf (CLCS) under the United Nations Convention on the Law of the Sea (UNCLOS) as a prime example of integrating geoscientific data and diplomatic negotiation to resolve issues like continental shelf extension. Raji and Awosika (2025) emphasize the crucial role of geoscientific diplomacy in maritime governance, transboundary resource management, disaster risk reduction, and climate change adaptation. They also provide practical success stories, including joint submissions and bilateral agreements, demonstrating how science can effectively inform policy decisions and promote stability and peaceful cooperation.

3   Conclusion

This diverse collection of chapters highlights the wide-ranging scope of geoethical considerations, spanning from the detailed nuances of research ethics to the expansive challenges of planetary governance and extraterrestrial exploration. By weaving together a rich variety of perspectives, the book reaffirms that geoethics is far more than an abstract philosophical idea; it is a practical and essential framework for responsible action across all geoscientific fields.

Ultimately, this volume serves as a compelling call to action, urging geoscientists, policymakers, and the broader public to adopt an integrated, ethically informed approach to our planet. It stresses that fostering interdisciplinary collaboration, valuing cultural diversity, and encouraging open dialogue are crucial for addressing complex global challenges and building a sustainable future where humanity can thrive in harmony with the Earth system. Through its comprehensive examination of geoethical principles and their real-world applications, this book offers both inspiration and practical tools to embed ethics at the core of all geoscience, ensuring that scientific progress truly serves the public good.

Acknowledgements We wish to sincerely thank the following colleagues for their valuable contributions as reviewers, providing insightful comments and suggestions that greatly enhanced the quality of the book's chapters: Mamoon Allan, Martin Bohle, Enrico Cameron, Valerio De Rubeis, Ross Dowling, Stan Finney, Jan Kunnas, Marco Pantaloni, Ester Stzein, and Greg Wessel.

References

Butler, R. W. H., Bond, C. E., & Cooper, M. (2025). Valuing diversity in geological interpretation: A warning from history. In S. Peppoloni & G. Di Capua (Eds.), *Geoethics and geosciences serving society—Reflections from the 37th international geological congress*. SpringerBriefs in Geoethics.

Greenbaum, D. (2025). Red planet, green ethics: Navigating the moral terrain of terraforming Mars. In S. Peppoloni & G. Di Capua (Eds.), *Geoethics and geosciences serving society—Reflections from the 37th international geological congress*. SpringerBriefs in Geoethics.

Hughes, M. W., & N. Kereikeepa (2025). Towards integrating indigeneity and geoethical practice in Aotearoa New Zealand. In S. Peppoloni & G. Di Capua (Eds.), *Geoethics and geosciences serving society—Reflections from the 37th international geological congress*. SpringerBriefs in Geoethics.

Kelman, I. (2025). Operational ethics for researching geohazards. In S. Peppoloni & G. Di Capua (Eds.), *Geoethics and geosciences serving society—Reflections from the 37th international geological congress*. SpringerBriefs in Geoethics.

Khan, S., Mandal, A., & Khan, T. N. (2025). Geoheritage, geotourism, and geoethics as support for rural development: a case study from Sundarbans Wetlands—A Ramsar site of West Bengal, India. In S. Peppoloni & G. Di Capua (Eds.), *Geoethics and geosciences serving society—Reflections from the 37th international geological congress*. SpringerBriefs in Geoethics.

Koster, E., Gibbard, P., & Maslin, M. (2025). The anthropocene event as a holistic foundation for earth governance. In S. Peppoloni & G. Di Capua (Eds.), *Geoethics and geosciences serving society—Reflections from the 37th international geological congress*. SpringerBriefs in Geoethics.

Peppoloni, S., Di Capua, G., & Bilham, N. (2025). The future challenges of geoethics: Navigating technology, sustainability, and social and professional responsibility. In S. Peppoloni & G. Di Capua (Eds.), *Geoethics and geosciences serving society—Reflections from the 37th international geological congress*. SpringerBriefs in Geoethics.

Raji, M., & Awosika, L. (2025). Geoscience diplomacy at the edge of the continental shelf. In S. Peppoloni & G. Di Capua (Eds.), *Geoethics and geosciences serving society—Reflections from the 37th international geological congress*. SpringerBriefs in Geoethics.

The Future Challenges of Geoethics: Navigating Technology, Sustainability, and Social and Professional Responsibility

Silvia Peppoloni⑩, Giuseppe Di Capua⑩, and Nic Bilham⑩

Abstract Geoethics is increasingly acknowledged as a critical field for examining the societal implications of geoscientific knowledge and its contribution to fostering a responsible and sustainable future for humanity. It engages with the ethical, cultural, and social dimensions of human interaction with the Earth system, focusing on natural resource use, land management, risk mitigation, and ecosystem protection. In the context of escalating global socio-ecological crises, geoethics demands more integrated and interdisciplinary approaches. These challenges necessitate a redefinition of the objectives, values, and ethical foundations underpinning scientific inquiry and technological innovation. Furthermore, contemporary governance frameworks, including decision-making and legislative processes, must be re-envisioned to adequately reflect the complexity and interconnectedness of an increasingly digitalized and globalized world. Central to the advancement of geoethics is the development and adoption of a methodological framework, conceptualized here as *Relational Geoscientific Pragmatism* (RGP), that advocates for pragmatic, context-sensitive solutions grounded in cross-disciplinary collaboration. It seeks to harmonize scientific understanding with ethical principles and societal values, drawing upon the philosophical orientation of ecological humanism. RGP emphasizes a relational perspective that affirms the intrinsic value of both humanity and the physical world, promoting a more holistic, inclusive, and ethically informed response to contemporary socio-environmental challenges. This chapter critically

S. Peppoloni (✉) · G. Di Capua
Istituto Nazionale di Geofisica e Vulcanologia, Rome, Italy
e-mail: silvia.peppoloni@ingv.it

International Association for Promoting Geoethics (IAPG), Rome, Italy

Chair on Geoethics of the International Council for Philosophy and Human Sciences (CIPSH-CG), Rome, Italy

Commission on Geoethics of the International Union of Geological Sciences (IUGS-CG), Rome, Italy

N. Bilham
Geoscience for Global Development (GfGD), London, UK

examines the principal challenges facing geoethics in the coming years and articulates how RGP, informed by ecological humanism, can serve as a guiding framework for both the scientific community and broader society. It aims to contribute to the ongoing discourse on sustainable development by advancing ethically robust pathways that integrate scientific knowledge with normative commitments to social and environmental justice.

Keywords Geoethics · Relational geoscientific pragmatism · Sustainability · Technology · Social responsibility · Future

1 Introduction: Global Anthropogenic Changes, Sustainability, and Ecological Humanism

In the context of growing concerns about climate change, the intensive exploitation of natural resources, pollution, and the spread of social inequalities and geopolitical tensions, human communities are experiencing an increasing sense of instability and fear for the future. This condition of uncertainty is further compounded by the gradual weakening of international governance bodies, which are increasingly perceived as incapable of effectively addressing major global challenges. The uncritical adoption of development models based solely on linear economic growth and technological advancement continues to significantly overlook the social, environmental, and cultural dimensions of human development (Cameron, 2023). In this scenario, rights, democracy, and ecological transition appear to be undergoing a phase of generalized contraction, highlighting the fragility of current paradigms and the urgent need for a profound rethinking of global priorities.

The dynamics between human societies and the environment continue to be shaped by outdated and unsustainable patterns, with little progress made in developing alternative visions that equitably integrate human well-being at all levels with ecological integrity. Assessments of the quality and health of a community remain closely tied to the concept of GDP (Gross Domestic Product), and international relations are still driven by archaic mechanisms based on power struggles for access to energy and mineral resources, in the absence of any commitment to a shared sense of rights. Moreover, the persistent reproduction of asymmetric power relations, based on a dominator–dominated logic, remains the primary mode of interaction shaping international balance. Sustainability, although now a globally recognized concept thanks in large part to the United Nations' promotion of the 17 Sustainable Development Goals (SDGs), often appears more aligned with commercial marketing strategies than with a genuine and substantial shift in economic paradigms (Agarwal et al., 2017). Finally, social distortions and ecological damage caused by the current system of relations among nation-states, global economy, and politics are piling up, further complicating efforts to address shared problems.

In this bleak and complex scenario, ethics reveals all its fragility in the attempt to contain the destructive impulses running through society. Yet it is precisely within this fragility that its strength lies: ethics gains renewed centrality as it calls us to take responsibility for choosing, today, the future we want, or are able, to build.

It is in this context that geoethical reflection finds its place, as it is called to engage with some of the most pressing challenges of our time (Peppoloni & Di Capua, 2022). Geo-environmental decision-making, from the extraction of natural resources to land use, from water management to disaster risk reduction, will have deep and lasting impacts on the planet's climate and ecosystems. However, these choices are not solely environmental in a narrow sense: they directly affect the human ecosystem, influencing fundamental aspects of collective life such as culture, economy, and politics, thereby helping shape the social and institutional dynamics of the future.

Geoethics is therefore called upon to grapple with the dilemma of how to balance economic and technological development and access to natural resources with the urgent need to eliminate or drastically reduce greenhouse gas emissions, contain pollution, preserve biodiversity, and reconcile the rights and development of local communities with the often-divergent needs of national and global societies.

A crucial aspect of sustainability concerns, in particular, the use of non-renewable resources. Mining, oil and gas extraction, and the exploitation of other exhaustible resources raise urgent questions about the extent to which, and the manner in which, it is ethically acceptable to extract and use such resources, and how we can accelerate, as swiftly as possible, the transition to an energy system that drastically reduces dependence on fossil fuels as the engine of industry and the economy (Smil, 2022).

Ultimately, geoethics is faced with a crucial question: how can we ensure that the use of natural resources does not compromise the ability of future generations to meet their own needs, nor substantially alter the habitability of the Earth system?

The solutions to be adopted must be grounded in the principles of justice and sustainability, both intra- and intergenerational, avoiding practices that risk irreversibly disrupting the planet's ecological balance. From this perspective, geoethics advocates for a model of development based on *Ecological Humanism*[1] (Peppoloni & Di Capua, 2023): a worldview that recognizes the interdependence between human

---

[1] Ecological Humanism is a broad and evolving interdisciplinary field, drawing from philosophy, environmental ethics, and social sciences. It advocates for environmental stewardship that is deeply informed by human values, cultural diversity, and ethical responsibility, to achieve a balanced relationship between humans and the natural world. The concept of "Ecological Humanism" emphasizes the integration of ecological awareness with human values and needs, as well as the interconnectedness of humans and nature, and advocates for a vision where human societies are consciously designed to enhance the well-being of both humanity and the environment. According to the work of René Dubos (1901–1982), key ideas of Ecological Humanism are the following: (a) Humans as Part of Nature (Humans are not separate from nature but an integral part of it. Understanding and respecting ecological systems is essential to human well-being); (b) Harmony Between Human and Environment (Humans must live in harmony with the environment, not through passive preservation, but through responsible stewardship that allows for both ecological balance and cultural expression); (c) Cultural and Local Adaptation (Dubos coined the phrase *Think globally, act locally,"* encouraging solutions that respect ecological principles while being tailored to local cultures and conditions); (d) Value of Human Experience (Unlike purely ecological perspectives that might sideline human needs, environmentalism must also serve human dignity, creativity, and fulfillment); (e)

beings and nature, and promotes a form of progress that respects both human needs and aspirations and the limits and requirements of ecosystems.

Table 1 presents the fundamental principles and defining features of ecological humanism, a vision emerging from a geoethical approach to the interaction between human beings and the Earth system. This vision is based on ecological and philosophical reflections developed in social ecology since around the 1970s (Morris, 2017).

In summary, ecological humanism represents a true paradigm shift from the traditional anthropocentric view, which still largely dominates our way of understanding the relationship between human beings and nature. This new approach promotes responsible coexistence among individuals, based on mutual respect and a deep awareness of the ecological responsibilities that bind us to the entire Earth system.

Table 1 Core principles and key characteristics of ecological humanism from a geoethical perspective

Tenet	Characteristics
Centrality of the human being as part of nature	Traditional anthropocentrism must give way to a new vision, in which the human being is no longer seen as a dominator of nature, but as an integral part of the Earth system and a responsible modifier agent
Harmony between human culture and nature	Human culture must integrate harmoniously with nature by adopting practices that respect ecosystems and abandoning those activities that severely compromise their dynamics
Ethical responsibility toward the Earth system	Human beings have the duty to care for socio-ecological systems by making responsible economic, social, and political choices and promoting a form of human development that respects the planet's limits and dynamics
Responsible development, intergenerational justice, and planetary governance	Human development must take into account the needs of future generations, ensuring equity and access to natural resources, and promoting the principles of global governance and planetary geo-citizenship
Geo-environmental education and awareness	Society must be based on the education of a citizenry with critical awareness, capable of recognizing the dignity of all entities that make up the Earth system and of promoting responsible social and environmental behaviors

---

Moral Responsibility (humanity has a moral obligation to care for the Earth, not only for survival but to foster a richer, more humane civilization). For further insights into this concept, as well as reference to some pioneering thinkers on Ecological Humanism, see Morris (2017).

2 Foundations of Geoethics

Geoethics promotes the vision of ecological humanism, rejecting the idea of the human being as a dominator of nature, and instead fostering responsibility and care for the Earth system.

Geoethics is founded on an awareness that humanity is increasingly influencing the Earth system (Ripple et al., 2024), and hence emerges as an ethical domain dedicated to critical analysis of the complex interactions between human beings and the planet (Peppoloni & Di Capua, 2022). It arises from the need to integrate scientific knowledge with the principles of individual and collective responsibility, and to reflect on the ethical implications of choices that might otherwise be considered purely scientific, technical, or political. It embraces multidisciplinary and interdisciplinary approaches, promoting a holistic and sustainable vision of the relationship between human activities and natural processes, for the common good and the protection of future generations.

Geoethics is a field of research and reflection dedicated to exploring and implementing the principles and values that guide appropriate behaviors and practices in all contexts where human activities interact with the Earth system (Peppoloni & Di Capua, 2022). Originally developed within the geosciences, geoethics initially aimed to raise awareness among geoscientists about their cultural and social roles, recognizing the ethical responsibility inherent in their scientific work (Peppoloni et al., 2019).

Over time, however, its scope has progressively broadened, extending beyond the scientific and professional community to promote a critical and informed attitude, rooted in science, for the benefit of society as a whole (Peppoloni & Di Capua, 2024).

From this perspective, geoethics offers cultural categories and behavioral values grounded in science, not by imposing prescriptive rules, but by encouraging responsible, informed action oriented toward the common good. Precisely because of its ability to integrate knowledge, ethics, and responsibility, geoethics has been proposed as a new global ethics (Peppoloni & Di Capua, 2020).

At the heart of geoethics is the human being, considered both individually and as part of a social community. Geoethics is conceived as a virtue ethics, focused on the human agent and the development of moral qualities such as prudence, integrity, and foresight. At the same time, it is proposed as an ethics of responsibility, which requires deep awareness of the consequences of one's actions within a complex system of relationships between humans and the environment, of which the human is an integral part rather than a dominator. Responsibility thus becomes the guiding criterion of informed, conscious individual action (Peppoloni & Di Capua, 2022).

Although it is not a prescriptive discipline, geoethics aims to identify and promote a set of shared principles and values, to which geoscientists and geocitizens voluntarily adhere. Its conceptual framework is based on three fundamental principles: dignity, freedom, and responsibility. These are accompanied by three aspirational principles: awareness, justice, and respect. From these principles arise a range of values that guide ethical action across the various domains of geoethics, from the individual to interpersonal and professional spheres, up to the social and environmental dimensions (Peppoloni & Di Capua, 2021).

These values are not prescriptive; rather, they guide decisions and actions that take into account the specific context in which interactions between human beings and the natural system occur. Contextualizing operational choices is essential to addressing the complexity of physical and social realities, recognizing that similar issues may require different responses depending on local specificities (Peppoloni et al., 2019).

The scientific knowledge provided by the geosciences represents a fundamental element of geoethics, as it offers geoscientists, other disciplinary specialists, decision-makers, and an informed citizenry essential tools to understand the interconnections that govern the Earth system, and to consciously address anthropogenic changes on a global scale in an informed and sustainable way. In this context, geo-education plays a crucial role in fostering environmental awareness, stimulating ethical reflection, and developing critical thinking (Bobrowsky et al., 2017).

Geoethics thus emerges as both a pedagogical and political project (Peppoloni & Di Capua, 2021), aimed at redefining the relationship between humanity and the Earth system through a synthesis of science and ethics, nature and culture, and multiple ethical perspectives—anthropocentric, biocentric, ecocentric, and geocentric (Peppoloni & Di Capua, 2023). Its ultimate goal is to overcome selfish individualism by promoting a conscious and compassionate planetary citizenship (Peppoloni & Di Capua, 2024).

3 Relational Geoscientific Pragmatism: A Methodological Framework for Geoethics

In an increasingly interconnected and vulnerable world, the integration of pragmatism and relational thinking within the geosciences is essential for addressing socio-environmental challenges ethically and responsibly. Building on work in recent years to broaden and deepen the theoretical underpinning and practical application of geoethics, *Relational Geoscientific Pragmatism* (RGP) is proposed as an overarching methodological framework for geoethical reflection, dialogue, decision-making, and action, operationalizing universal principles and shared values in a context-sensitive way.

The RPG framework advocates a holistic approach in which geosciences are intrinsically connected to social and environmental responsibility and sustainability, offering a way to navigate the complexities of geoscientific practice in the context of contemporary societies based on respect for the environment, sustainable resource management, and the well-being of present and future generations. Central to this approach is the pursuit of context-specific solutions, underpinned by rigorous scientific methodologies that assess the dynamic relationships between natural phenomena, societal needs, and decision-making processes. The RPG framework can also help address the key challenges now facing geoethics, and frame a call to action to the geoscience community and beyond.

Table 2 maps the key characteristics of geoethical methodology to the RPG framework

These categories and characteristics should prompt questions when tackling any geoethical challenge and inform their context-specific answers:

- What geoscientific knowledge, data, analysis and expertise are of value in the situation at hand?
- What other scholarly and professional fields should inform decision-making and action?
- Who are the relevant stakeholders and communities?
- What values and principles are relevant, reflecting stakeholder perspectives and concerns?
- How compatible are they?
- What ethical issues, dilemmas and choices arise from seeking to integrate or balance these values?
- What possible scenarios could be envisioned to address these issues and/or dilemmas?
- What are the processes through which stakeholder engagement and dialogue should take place?
- How are shared decisions to be made, and action taken, and by whom?

The RGP framework, then, is not a prescriptive method, but a structured values-driven pathway for responding responsibly to geoethical challenges, rooted in context. It can guide decision-makers and other stakeholders through contemporary challenges, and support the development of geo-governance models that balance scientific rigor with ethical and societal considerations.

4   The Challenges of Geoethics

Geoethics provides a coherent and integrated framework, linking scientific, social, cultural, and ethical aspects that enables a systematic and holistic approach to urgent, complex, interconnected challenges arising from anthropogenic global changes, including climate change, natural resource management, and the reduction of social inequalities.

There are a number of fundamental challenges to operationalizing this framework. A delicate balance must be struck between the universality of its ethical principles and the need to respond to the specificities of local socio-cultural contexts. Moreover, geoethics must deal with the difficulty of operating under conditions of uncertainty, where scientific predictions are not always conclusive, and solutions must be flexible, adaptable, and resilient.

Table 2 *Relational Geoscientific Pragmatism* as a framework for geoethical methodology

Category	Characteristic	Description
Geoscience	Knowledge and data	Geoscientific knowledge and objective, verifiable, up-to-date data are essential requirements for understanding natural phenomena and environmental dynamics, and enabling informed and evidence-based choices. Such knowledge is vital to challenges related to resource management, climate change, environmental sustainability and disaster risk reduction
	Scientific analysis	The application of rigorous and responsible scientific analysis to geoscientific knowledge and data is also essential. It informs decision-making by providing authoritative context-specific information, reliable risk assessments, and realistic future scenarios. By placing scientific analysis at the core of decision-making, the geoethics framework provides a structured means to navigate value conflicts and ethical dilemmas that inevitably arise in decision-making processes
	Professional expertise and judgment	The role of geoscientists goes beyond provision of data and analysis, and they are often required to exercise professional expertise and judgment. Geoethics calls on all stakeholders to act responsibly, demonstrating awareness and accountability, and carefully considering the consequences of their actions while balancing diverse and often competing interests. Professional standards play an important role in underpinning and assuring geoscientists' individual and collective responsible and ethical behavior, underpinned by ethical codes and accreditation schemes operated by professional bodies
Relationality	Interdisciplinarity	An interdisciplinary approach fosters a holistic understanding of both natural systems and societal dynamics. Environmental challenges are inherently complex and demand the integration of diverse fields, including geosciences, social sciences, economics, law and philosophy. Interdisciplinarity can be framed as relationality between disciplines and professions. A key responsibility for geoscientists is to recognize and nurture relations between geoscience and other relevant disciplines and perspectives, providing a context-sensitive structure and granularity to interdisciplinarity as applied to geoethical challenges

(continued)

Table 2 (continued)

Category	Characteristic	Description
	Social justice	Relationality to communities and individuals, especially marginalized ones, places social justice at the heart of addressing geoethical challenges. It is incumbent on all stakeholders, especially those with most power, to exercise social responsibility. Awareness of and sensitivity to marginalized and disempowered communities is also essential to identifying relevant stakeholders whose voices should be heard
	Sustainability	Geoethical approaches emphasize consideration and elucidation of relationality to future generations, not just current ones, as well as to non-human realms, fostering environmental and intergenerational responsibility
Pragmatism	Integrating universal values with context-specific factors	Geoethics is action- and solution-oriented, but values-driven. It is a practical endeavor aimed at making positive change in the world. Its practice is context-sensitive and its methods and solutions are context-dependent. However, this should not be mistaken for mere relativism. Core principles and values have driven geoethics' development, and continue to frame and guide its practice. Effective applied geoethics requires context-sensitive interpretation and balancing of principles and values in practical situations, in collaboration with relevant stakeholders and communities
	Openness, dialogue, inclusivity and accessibility	Geoethics advocates for the active involvment of all relevant stakeholders, including scientists, policymakers and local communities, in decision-making processes. This underpins integration and alignment of diverse values and perspectives, ensuring that proposed solutions are assessed not only for their technical feasibility but also for their ability to meet the needs and expectations of all parties involved, fostering a dynamic balance between ecological integrity and social well-being. In order to be equitable and effective, engagement and dialogue among stakeholders and communities must be transparent, inclusive and accessible—including through effective communication of geoscience and other specialist knowledge and information

(continued)

Table 2 (continued)

Category	Characteristic	Description
	Defining and exploring ethical dilemmas and scenarios	At the heart of geoethical methodology is the identification and critical analysis of ethical dilemmas arising from the relationship between humans and the environment, such as balancing economic development with environmental conservation, ensuring intergenerational justice, and protecting vulnerable communities. Consideration of the factors set out in this framework facilitates the creation of scenarios, by anticipating the potential outcomes of actions, and evaluating them through the lenses of sustainability, equity and environmental integrity

Another challenge lies in overcoming political and social resistance to the adoption of sustainable and ethical policies, as well as in defining a clear and recognized role for geoethics within the geosciences, so that it can meaningfully contribute to professional training and interdisciplinary dialogue.

At the same time, geoethics must be able to translate its core principles into concrete actions in the educational and political spheres. These actions should inspire both individual and collective behaviors oriented toward care and respect for the planet, generating a positive impact that extends beyond academic and professional spheres.

In the following subsections, the main challenges of geoethics are described, with the aim of shaping a call to action in the geoscience community and beyond, and ultimately promoting a shift in human action toward a culture of respect, responsibility, and care for one's vital niche.

4.1 Operationalising Geoethics

4.1.1 Balancing Universal Principles and Contextual Applications

Imposing prescriptive norms or operational guidelines without adequate consideration of the contextual framework can lead to several critical risks, including:

- Generating resistance and rejection: Local communities may strongly oppose and reject externally imposed measures that do not reflect their needs or socio-cultural realities.
- Inapplicability: Guidelines that are too generic or decontextualized may prove ineffective or difficult to implement in local settings characterized by specific challenges or unique features.

- Loss of legitimacy: Prescriptions that fail to account for local particularities may be seen as top-down impositions, diminishing their legitimacy and weakening the sense of shared responsibility.
- Unintended side effects: The absence of thorough contextual analysis can lead to unforeseen consequences, as adopted solutions may not align with the complexity of local dynamics and could generate new problems.
- Undermining social cohesion: Norms and orientations perceived as foreign to local cultures may spark internal or inter-community conflicts, damaging social cohesion and hindering constructive dialogue.

On the other hand, there are also risks in becoming unanchored from the universal principles, values, and theoretical frameworks of geoethics, including lack of credibility in some academic and political contexts, weakening of ethical foundations, and expediency at the expense of truly responsible action. A key challenge is thus to manage the potential relativism inherent in the concept of contextualization. The goal is to develop ethical orientations that are attuned to local specificities, responsive to global imperatives, and founded on universal values and principles (Peppoloni & Di Capua, 2023).

4.1.2 Dealing with Decision-Making Under Conditions of Uncertainty

The management of natural phenomena and their associated risks in the context of varying levels of scientific uncertainty requires effective communication of concepts such as uncertainty, risk, and probability to non-specialist stakeholders, and the development of shared understandings of these concepts across disciplines to inform risk mitigation strategies and sustainable and responsible decision-making.

4.2 Overcoming Barriers

4.2.1 Achieving Societal Acceptance and Engagement

One of the main challenges for geoethics is overcoming the resistance or apathy of many individuals and social groups when facing complex and uncertain future scenarios. It will be essential to raise awareness beyond the scientific community and engage society more broadly with the principles and values of geoethics, and the vision of ecological humanism, through a wide range of communication tools and collaborations with other forms of human expression, such as art.

4.2.2 Overcoming Political and Social Resistance

A crucial challenge will be to develop effective strategies to promote the principles and values of geoethics within political and social contexts that resist changing perspectives regarding individual, social, and geopolitical relationships. The political project of geoethics will inevitably come into conflict with entrenched powers—autocracies, theocracies, and oligarchies, but also embedded economic and political interests—that may hinder the adoption of a new ethical paradigm. The geoethical vision of planetary geo-citizenship implies a profound transformation of current systems of community, national, and regional relations, and of transnational geo-governance structures for the Earth system.

4.2.3 Maintaining Stability During Transformation

Although geoethics advocates for radical changes in the way we think and act, one of its main challenges is implementing such transformations without triggering a generalized regression of society, widespread backlash, or even a return to less inclusive or more authoritarian forms of social organization. The perception of imposed change in the absence of an effective strategy, community involvement, and consensus may spark identity-driven resistance and rejection, fueling political and social polarization. To avoid these risks, transformative processes must take into account the following elements:

- The active involvement of communities, in collaboration with institutions and the scientific community.
- The valorization of local knowledge as a resource for shared transformation.
- Raising awareness and knowledge of geoethical culture through education and public engagement.
- The training of leaders and professionals capable of translating geoethical values into concrete actions.
- Mediating between innovation and tradition, integrating societal change and new behavioral models with traditional practices, acquired rights and social achievements.
- Monitoring and evaluating the effects of transformations, adapting them promptly to evolving social, economic and environmental conditions.

4.3 Extending the Reach and Impact of Geoethics

4.3.1 Geoethics for Policy-Making

The framework of geoethical principles, values, and methods outlined in this essay and elsewhere provides a fundamental foundation for the development of policies and legislative tools capable of reshaping contemporary social and political dynamics, guiding decision-making at global, national, and local levels. Geoethics not only encourages responsibility in scientific and professional practices, but also aims to act as a catalyst for change in public policy, in a wide range of fields such as environmental protection, modern slavery, and human rights, and requirements for responsible resource supply chains.

In this context, activism plays a crucial role, complementing the professional actions of geoscientists. On one hand, geoscience professionals are called to provide data, analyses, and technical solutions for managing environmental risks and promoting sustainability. On the other hand, their active engagement, rooted in knowledge and expertise, can drive change by raising collective awareness, engaging communities, and putting pressure on governments and corporations to adopt fairer and more ethical practices and policies.

The interaction between geoscientists' technical expertise and demands emerging from civil society can create a powerful alliance capable of positively influencing political, economic, and social decisions. Such a synergistic approach, combining science, ethics, and civic engagement, is essential to addressing the global challenges of our time and steering societies toward a sustainable and inclusive future.

4.3.2 Developing an Effective Pedagogy

Translating the theoretical framework of geoethics into effective educational and engagement programs, targeted at both future generations and current leaders and decision-makers, represents one of the most challenging tasks for geoethics. This process requires the development of appropriate teaching resources and methodologies that integrate solid geoscientific knowledge with critical reflection on the cultural and social value of geosciences. It is essential to promote an education that stimulates critical thinking, fosters ethical responsibility, and raises awareness of socio-environmental risks.

4.3.3 Strengthening the Role of Geoethics in Geosciences

It is essential to fully integrate geoethics into both the research and practice of geosciences. This means clarifying how geoethics can enrich and guide the work of geoscientists, providing an ethical framework that not only supports their activities but also amplifies their social impact. Tools for achieving this shift include institutional partnerships, education and training (including through continuing professional development), and professional accreditation standards that meet evolving societal needs.

4.3.4 Promoting Interdisciplinary Collaboration

Geoethics, by its nature, is an inherently interdisciplinary field that requires an integrated approach across different areas of knowledge. One of the key challenges will be to develop increasingly effective forms of collaboration, fostering constructive dialogue between disciplines such as geosciences, economics, sociology, law, and the humanities. Overcoming barriers between disciplinary languages and methods is essential to addressing complex problems in a holistic way. Interdisciplinarity must also promote the integration of different cultural traditions and worldviews, valuing not only scientific perspectives but also indigenous and local knowledge that preserves ancestral approaches to land management, resources, and risks. However, while it is crucial to open up to a plurality of perspectives, scientific methodology must remain the core of the decision-making process: only through verifiable scenarios, reliable data, and rational analysis can effective solutions be built. The challenge lies in reconciling the robustness of the scientific approach with the richness of intercultural and interepistemic dialogue, capable of developing truly sustainable, inclusive, and shared visions for the future.

4.4 Envisioning a Geoethical Future

4.4.1 Geoethics for Emerging Complex Global Issues

Geoethics must address emerging technologies, such as artificial intelligence (Cleverley, 2024), and activities that involve significant geo-environmental impacts, such as climate geo-engineering and deep-sea mining (Peppoloni & Di Capua, 2020). These practices, whose systemic risks are not yet fully understood, raise important ethical and operational questions, as their long-term effects could result in severe and unpredictable consequences for the environment and the prospects of generations to come. Therefore, it will be essential to develop robust yet flexible ethical frameworks capable of addressing these issues in a dynamic and contextual manner. Such guidelines must consider not only the intrinsic scientific uncertainties, but also the objective challenges related to the timely assessment of the impacts of rapidly evolving technologies and the concomitant development of governance structures. In the face of accelerating technological innovation, geoethics must promote a continuous process of ethical oversight and updating to ensure that decisions are based on comprehensive, up-to-date evaluations grounded in the best available evidence.

4.4.2 Promoting Geoethics as a New Horizon for Global Ethics

This represents a fundamental challenge, requiring a careful comparative analysis between geoethical thought and the ethical systems and values of different human cultures. The goal of such a comparison is to identify areas of convergence that, while fully respecting cultural diversity and local traditions, can form a shared foundation for building a responsible planetary citizenship. This citizenship would not only be aware of its rights, but also deeply conscious of its duties toward other living species and the natural entities that make up the Earth system.

5 Rethinking Geosciences in Light of Geoethics

The challenges of geoethics are closely linked to perceptions of the often-underestimated social role of geosciences. The latter play an essential role in understanding natural phenomena, land management, and the analysis of environmental and climate risks (Tewksbury, 2017), as well as in achieving the 17 UN SDGs (Senger, 2024). Despite this, the perceived social usefulness of geosciences remains surprisingly low, particularly in Global North countries (Rogers et al., 2024).

The reasons for this disconnection are complex and go well beyond a simple lack of awareness. Although we do not delve into the analysis of these causes in this essay, we highlight two aspects that we consider particularly relevant.

First, the geosciences have provided a deep understanding of natural phenomena and their impacts on the environment and society. However, many of the discoveries and analyses produced in this field are not communicated effectively, significantly reducing the potential contribution of these disciplines to public policy development and the transformation of collective behaviors (Stewart & Hurth, 2021).

Second, deeply ingrained cultural assumptions perpetuate a distorted perception of the role of geosciences.

On the one hand, they are often not recognized as an integral part of the "technical" sciences, but rather as abstract and distant disciplines, difficult to connect to the concrete everyday needs of people (Bohle, 2015). In a world increasingly dominated by technology and the search for quick solutions, applied sciences tend to be favored for their apparent ability to offer clear and immediate answers to urgent problems, and technology is valued for its ability to produce certain results based on structured, easily interpretable, and unequivocal models, such as the binary logic of computing or the precision of engineering processes.

On the other hand, the geosciences are not perceived to conform to the model of so-called "pure" natural sciences, such as physics or chemistry, which tend to be more readily understood and valued, as they are seen as based on universal laws and verifiable predictions, generating a sense of control and reliability.

In contrast with both these received models of "useful" sciences, geosciences operate in a domain dominated by uncertainty and probability (Hetényi et al., 2022). Their primary goal is not to offer immediate or deterministic solutions, but rather to understand and analyze extremely complex natural phenomena that are not easily amenable to simple or definitive answers. Geosciences often work with probabilistic scenarios, attempting to model the evolution of events such as earthquakes and volcanic eruptions, climate change, the occurrence of natural resources, and the response of the geosphere to geo-engineering activities—phenomena influenced by a multitude of variables, many of which are difficult to identify, measure, or predict with accuracy. The uncertain and probabilistic nature of geoscientific analyses appears less compatible with the demands of contemporary political, economic, and social systems that favor knowledge based on immediate certainties and tangible results, even if such apparent certainty is spurious.

Despite addressing crucial practical issues, and having both firm theoretical foundations and powerful analytical and predictive value, geosciences may thus be perceived as less relevant in the context of practical solutions. This reductive perception contributes to limiting their social and political impact, just when a systemic and integrated vision of the Earth is more necessary than ever. It is precisely the ability of geosciences to analyze complex systems and their dynamics, anticipate intricate evolutionary scenarios, and develop cognitive tools for managing long-term risks that makes them essential tools for understanding and addressing the major global challenges of our time.

The task of geoethics, therefore, is not limited to strengthening the link between geosciences and society (Stewart, 2023, 2024), but also consists in bridging the perceptual gap that still separates these disciplines from the social fabric, while simultaneously contributing to a critical reflection on the historical and current role of geosciences in shaping the present socio-ecological crisis, and the particular responsibilities this entails (Leinfelder et al., 2024; Peppoloni & Di Capua, 2024). In this perspective, geoethics must contribute to the development and promotion of modes of education, culture, and scientific communication that clearly and accessibly convey the value of uncertainty and the power of geosciences to help understand and harness it. Uncertainty should not be perceived as a limitation, but rather as an integral part of knowledge that engages with the dynamic reality of natural and social systems. In a world characterized by objective and growing uncertainty, geoethics can play a fundamental role in fostering a closer integration between science, politics, and society, demonstrating that the great global challenges cannot be addressed with simplistic or rigid solutions, but require a conscious, informed approach open to the complexity and uncertainty of the real world.

6 Conclusions: Three Urgent Priorities for Applied Geoethics

In the context of complex global socio-environmental challenges, geoethics emerges as a fundamental tool for developing approaches that can integrate technological innovation, ethical responsibility, and social justice. Achieving this objective, and meeting the challenges facing geoethics set out above, will require significant work. A fully developed program to address these challenges, guided by the RGP framework, is beyond the scope of this essay. In this concluding section, therefore, we present three emblematic examples that highlight the concrete application of geoethics in particularly relevant areas and that require urgent action. These examples represent a call to action to geoscientists and others seeking to effect a geoethical transition, not only to address the specific issues raised but also to join in the project of developing a wider program of transformation, guided by a shared framework.

6.1 Advanced Technologies, Environmental Monitoring, and Data Access

The rapid evolution of geospatial technologies, including high-resolution satellites, drones, remote sensors, and real-time monitoring systems, has radically transformed the capabilities of the geosciences, offering unprecedented access to detailed, dynamic, and continuous environmental data. These tools represent an extraordinary resource for understanding and addressing complex phenomena such as climate change, ecosystem degradation, and natural hazards, significantly improving strategies for prevention, response, and planning.

However, the use of these technologies raises significant ethical questions (Bennett et al., 2023), particularly regarding privacy, surveillance, and the management of sensitive data. Geospatial information can be used not only for scientific and environmental purposes but also to monitor individual movements, economic or social activities, and to collect data on private or culturally sensitive territories. This implies delicate issues related to informed consent, transparency in the use of information, and the protection of individual and collective rights.

In this context, geoethics is called upon to define guiding principles and clear limits for the responsible use of geospatial technologies, ensuring that the benefits derived from innovation do not compromise human dignity or violate the autonomy of local communities. It is essential now to develop ethical codes of conduct, policies for equitable and responsible data access (Cocco et al., 2025), and participatory decision-making mechanisms, so that technological innovation aligns with an ethical vision focused on justice, transparency, and sustainability.

6.2 Environmental and Social Justice

Decisions in the geo-environmental field have consequences that are not distributed equally; often, the most severe impacts fall on the most vulnerable populations. Indigenous and low-income communities, and socially marginalized groups disproportionately suffer the consequences of extractive activities, the construction of infrastructure in high-risk areas, such as seismic zones or coastlines exposed to extreme events, and the effects of climate change, despite having contributed minimally to the causes of these problems (Scheidel et al., 2023).

In this context, geoethics is called upon to advocate for the principles of environmental, distributive, and epistemic justice, promoting fair, inclusive decision-making processes that protect the rights of the most exposed communities (Bobadilla et al., 2024). Geoethical practices must include concrete mechanisms for the active participation of local populations, ensuring that land and natural resource management is based on principles of equity, transparency, and responsibility. Only through an authentically participatory and democratic approach will it be possible to build a more just, resilient, and sustainable relationship between human society and the environment, valuing the knowledge of communities and laying the foundations for a conscious ecological citizenship.

6.3 Geoethical Education and Awareness

A core strategic and transformative dimension of geoethics is education and training, understood not only as the transmission of knowledge but also as a cultural tool aimed at building a shared ethical consciousness. Decisions in the geoscientific domain can no longer be the exclusive prerogative of experts, as their environmental, social, and cultural implications affect society as a whole. For this reason, it is essential to empower the broader public with the knowledge and tools needed to understand and critically engage with issues of sustainability, environmental justice, and the complex relationship between humans and the Earth system. A true cultural shift requires the promotion of values such as respect for the environment, intergenerational equity, scientific responsibility, and dialogue between different forms of knowledge.

In this process, universities, research centers, and other educational institutions play a key role: It is necessary to integrate interdisciplinary programs into curricula that combine geosciences with ethics, social sciences, and sustainability. Only by training professionals who can merge scientific rigor with social awareness will it be possible to steer future decisions toward more equitable, inclusive, and forward-thinking models of development, thereby contributing to the creation of a responsible, active, informed, and truly conscious planetary citizenry, capable of meaningfully participating in decision-making processes that affect the planet's future.

References

Agarwal, N., Gneiting, U., & Mhlanga, R. (2017). Raising the bar: Rethinking the role of business in the sustainable development goals. *Oxfam Discussion Paper*, Retrieved April 7, 2025, from https://www-cdn.oxfam.org/s3fs-public/dp-raising-the-bar-business-sdgs-130217-en_0.pdf

Bennett, M. M., Gleason, C. J., Tellman, B., Alvarez Leon, L. F., Friedrich, H. K., et al. (2023). Bringing satellites down to Earth: Six steps to more ethical remote sensing. *Global Environmental Change Advances, 2*. https://doi.org/10.1016/j.gecadv.2023.100003

Bobadilla, H., Di Capua, G., Hesselbein, C., Peppoloni, S., & Lampis, F. (2024). Advancing epistemic justice with local knowledge: A process indicator for EU climate adaptation policymaking. In E. Galende Sánchez, A.H. Sorman, V. Cabello, S. Heidenreich, & C.A. Klöckner (Eds.), *Strengthening European climate policy* (pp. 49–60). Palgrave Macmillan. https://doi.org/10.1007/978-3-031-72055-0_5

Bobrowsky, P., Cronin, V., Di Capua, G., Kieffer, S., & Peppoloni, S. (2017). The emerging field of geoethics. In L.C. Gundersen (Ed.), *Scientific integrity and ethics: With applications to the geosciences* (pp. 175–212). Special Publication American Geophysical Union. Wiley. https://doi.org/10.1002/9781119067825.ch11

Bohle, M. (2015). Simple geoethics: An essay on daily Earth science. In S. Peppoloni & G. Di Capua (Eds.), *Geoethics: The role and responsibility of geoscientists*. Geological society (Vol. 419, pp. 5–12). Special Publications. https://doi.org/10.1144/SP419

Cameron, E. (2023). Manfred Max Neef's human scale development and geoethics. *Journal of Geoethics and Social Geosciences, 1*(1), 1–25. https://doi.org/10.13127/jgsg-28

Cleverley, P. H. (2024). Ethical recommendations for artificial intelligence technology in the geological sciences—With a focus on language models. *Journal of Geoethics and Social Geosciences, 1*(2), 1–25. https://doi.org/10.13127/jgsg-63

Cocco, M., Paciello, R., Bailo, D., Locati, M., Tanlongo, F., et al. (2025). The ethical dimension of sharing solid Earth Science data. *Journal of Geoethics and Social Geosciences, 2*(Special Issue), 1–30. https://doi.org/10.13127/jgsg-64

Hetényi, G., Balázs, L., Barcza, Z., Békési, E., Győ, E., et al. (2022). The inherent uncertainty in geosciences. *Acta Geodaetica Et Geophysica, 57*, 411–418. https://doi.org/10.1007/s40328-022-00392-6

Leinfelder, R., Thomas, J. A., Vidas, D., Williams, M., & Zalasiewicz, J. (2024). Geoethics and the anthropocene: Five perspectives. In S. Peppoloni & G. Di Capua (Eds.), *Geoethics for the future: Facing global challenges* (pp. 69–83). Elsevier. https://doi.org/10.1016/B978-0-443-15654-0.00005-0

Morris, B. (2017). *Pioneers of ecological humanism: Mumford, Dubos and Bookchin*. Black Rose Books. Retrieved June 10, 2025, from https://files.libcom.org/files/Pioneers_of_Ecological_Humanism__PDF_.pdf

Peppoloni, S., Bilham, N., & Di Capua, G. (2019). Contemporary geoethics within geosciences. In M. Bohle (Ed.), *Exploring geoethics: Ethical implications, societal contexts, and professional obligations of the geosciences* (pp. 25–70). Palgrave Macmillan. https://doi.org/10.1007/978-3-030-12010-8_2

Peppoloni, S., & Di Capua, G. (2020). Geoethics as global ethics to face grand challenges for humanity. In G. Di Capua, P.T. Bobrowsky, S.W. Kieffer, & C. Palinkas (Eds.), *Geoethics: Status and future perspectives*. Geological society (Vol. 508, pp. 13–29). Special Publications. https://doi.org/10.1144/SP508-2020-146

Peppoloni, S., & Di Capua, G. (2021). Geoethics to start up a pedagogical and political path towards future sustainable societies. *Sustainability, 13*(18), 10024. https://doi.org/10.3390/su131810024

Peppoloni, S., & Di Capua, G. (2022). *Geoethics: Manifesto for an ethics of responsibility towards the Earth*. Springer. https://doi.org/10.1007/978-3-030-98044-3

Peppoloni, S., & Di Capua, G. (2023). Geoethics for redefining human-earth system Nexus. In G. Di Capua & L. Oosterbeek (Eds.), *Bridges to global ethics: Geoethics at the confluence of humanities and sciences* (pp. 5–23). SpringerBriefs in Geoethics. Springer. https://doi.org/10.1007/978-3-031-22223-8_2

Peppoloni, S., & Di Capua, G. (2024). Etiology of the ecological crisis: Building new perspectives for human progress through geoethics. In S. Peppoloni & G. Di Capua (Eds.), *Geoethics for the future: Facing global challenges* (pp. 51–67). Elsevier. https://doi.org/10.1016/B978-0-443-15654-0.00009-8

Ripple, W. J., Wolf, C., Gregg, J. W., Rockström, J., Mann, M. E., et al. (2024). The 2024 state of the climate report: Perilous times on planet Earth. *BioScience, 74*(12), 812–824. https://doi.org/10.1093/biosci/biae087

Rogers, S. L., Giles, S., Dowey, N., Greene, S. E., Bhatia, R., et al. (2024). "you just look at rocks, and have beards" Perceptions of geology from the United Kingdom: A qualitative analysis from an online survey. *Earth Science, Systems and Society, 4.* https://doi.org/10.3389/esss.2024.10078

Scheidel, A., Fernández-Llamazares, Á., Bara, A. H., Del Bene, D., David-Chavez, D. M., et al. (2023). Global impacts of extractive and industrial development projects on Indigenous Peoples' lifeways, lands, and rights. *Science Advances, 9*(23). https://doi.org/10.1126/sciadv.ade9557.

Senger, K. (2024). Sustainable development goals and the geosciences: A review. *Earth Science, Systems and Society, 4.* https://doi.org/10.3389/esss.2024.10124

Smil, V. (2022). Come funziona davvero il mondo – Energia, cibo, ambiente, materie prime: le riposte della scienza (Translation in Italian of the book "How the World Really Works", 2021, Penguin). Giulio Einaudi editore, Torino.

Stewart, I. S., & Hurth, V. (2021). Selling planet Earth: Re-purposing geoscience communications. In G. Di Capua, P.T. Bobrowsky, S.W. Kieffer, & C. Palinkas (Eds.), *Geoethics: Status and future perspectives.* Geological society (Vol. 508, pp. 265–283). Special Publications. https://doi.org/10.1144/SP508-2020-101

Stewart, I. S. (2023). Geology for the wellbeing economy. *Nature Geoscience, 16*, 106–107. https://doi.org/10.1038/s41561-022-01110-1

Stewart, I. S. (2024). Geoscience for Earth stewardship, sustainability, and human well-being: A conceptual framework for integrating planet, prosperity, and people. In S. Peppoloni & G. Di Capua (Eds.), *Geoethics for the future: Facing global challenges* (pp. 173–189). Elsevier. https://doi.org/10.1016/B978-0-443-15654-0.00029-3

Tewksbury, B. (2017). Why geoscience? Why geoscience in the context of societally and culturally relevant questions? Retrieved June 10, 2025, from https://serc.carleton.edu/integrate/workshops/african-education/essay/179783.html

Valuing Diversity in Geological Interpretation: A Warning from History

Robert W. H. Butler⊙, Clare E. Bond⊙, and Mark A. Cooper

Abstract Assessing uncertainty in the geological understanding of the subsurface, as with many areas of applied science, demands seeking out and valuing alternative interpretations. But how inclusive are we collectively in promoting and valuing diverse interpretations? We explore this question using a parable drawn from the history of earth sciences—examining the conduct of the Continental Drift Debate of the 1920s. Aggressive, intemperate language by leading North America geologists devalued interpretations and the competence of geologists who proposed a tectonically mobile Earth. Our contention is that this reinforced groupthink that rejected alternative hypotheses and observations that conflicted with their own explanations, and argue that this in turn inhibited research on global tectonics for the decades that followed. We explore how increasing the diversity of participants in scientific debates along with treating diverse views with respect and tolerance could avoid these outcomes.

Keywords Diversity · Groupthink · Continental drift

1 Introduction

How well do earth scientists—or indeed natural scientists in general—value, respect and incorporate research and opinion that conflict with their own models and hypotheses?

R. W. H. Butler (✉) · C. E. Bond · M. A. Cooper
Fold-Thrust Research Group, Geology & Geophysics, School of Geosciences, University of Aberdeen, Aberdeen AB24 3UE, UK
e-mail: rob.butler@abdn.ac.uk

C. E. Bond
e-mail: clare.bond@abdn.ac.uk

M. A. Cooper
e-mail: mark@sherwoodgeo.com

© The Author(s), under exclusive license to Springer Nature Switzerland AG 2025
S. Peppoloni and G. Di Capua, *Geoethics and Geosciences Serving Society*,
SpringerBriefs in Geoethics, https://doi.org/10.1007/978-3-032-03754-1_3

This question lies at the heart of challenges facing applied earth science in supporting the energy transition and engineering in the earth's near subsurface that this necessitates. As in all areas of human investigation, earth scientists need to encourage, then evaluate diverse explanations if collectively we are to assess uncertainty in geological interpretations. Our aim here is to highlight the risks of not being open to diversity, both in terms of ideas and in valuing the contributions of other communities of scholars. We recognise that looking at modern science and scholars is difficult—analysis of bias and decision-making risks alienating scholars by being accusatory, and that is unlikely to promote reflection by those being accused. This can further entrench non-inclusive behaviours. In contrast, historical scientific debates and controversy provide rich resources for exploring how these behaviours play out. Here we draw lessons from the much-chronicled rejection of continental drift by North American earth scientists in the 1920s.

There are numerous histographies and commentaries that address the debates around continental drift in the 1920s, centred on the publication by Alfred Wegener of his *Die Entstehung der Kontinente und Ozeane* in its third edition in 1922 and its English translation (*The origin of continents and oceans*) in 1924. It is not our intention to rehearse these in detail here—readers are referred to accounts by Frankel (1985, 2012), Newman (1995) and Oreskes (1999) amongst others. Romano et al. (2017) add reflections on the reception of Wegener's ideas in Italy. Here we focus on the conduct of the debate. We propose that the language used by those opposed to Wegener's theory served to intimidate dissenters and, with a legacy lasting for several decades, contributed to a climate that strongly discouraged new research on global tectonics. After describing the science and conduct of a historical debate, we reflect on possible learnings.

2 The Continental Drift Debate: Fixism Versus Mobilism and the Players

Famously, in November 1926, a group of leading American earth scientists met in New York at a symposium of the American Association of Petroleum Geologists. Their purpose was to debate Alfred Wegener's theory of continental drift—a tectonic theory that invoked large-scale mobility of the earth's surface charted by changes in the relative position of continents over geological time. The timing reflected the first publication of Wegener's "*Origin of Continents and Oceans*" in the English language (Wegener, 1924) and hence brought his proposition of a tectonically mobile earth to the widespread attention of geologists in North America. This event has passed into folklore as the pivot for the radical rejection by this geological community of continental drift. Rather, they asserted that the earth's surface had never experienced significant horizontal displacements—indeed the physical impossibility of such deformation (e.g. Le Grand, 1986).

Studies of the continental drift debate distil protagonists into two distinct camps—the "fixists" and the "mobilists". The terms relate to the positions of continents and oceans through geological time, as related to the formations of the world's mountain ranges. Earth science had to wait until the late 1960s before this "fixist" view of the planet was near-universally rejected—through the adoption of the all-embracing theories of plate tectonics.

Team "fixist" held that the continents and oceans had not changed and therefore that mountain ranges formed with only minor horizontal displacements. At the start of the 20th century, a chief advocate of this model was Thomas Chrowder Chamberlin (1843–1928), founding professor of geology at the University of Chicago. Nowadays TC Chamberlin is remembered for promoting the building of scientific knowledge through the assessment of multiple working hypotheses—in which evidence is sought then used to assess competing explanations. But Chamberlin was, during his lifetime, equally known for the development of "punctuated diastrophism"—whereby mountain ranges formed by sporadic contraction of the planet, with deformation focused on continental margins. It was a proposal grounded in the distribution of mountain ranges in North America. This diastrophism concept permitted only minor amounts of contraction and no significant lateral motions. The earth's continents and oceans were primary features formed in place as the planet accreted. Chamberlin published his ideas repeatedly in the publication he founded and edited—the Journal of Geology. His static view of fixed palaeogeography was adopted by other North American scholars, with Chamberlin's son, Rollin (1881–1948) being an especially outspoken advocate. Following his father's retirement in the early 1920s, he took over chairmanship of the department in Chicago and editorship of the Journal of Geology. He used these positions and status to support and perpetuate his father's theory with other earth scientists across North America (see commentary in Pettijohn, 1970, p. 96). They included Bailey Willis (1857–1949) and Charles Schuchert (1858–1942), both of whom served as presidents of the Geological Society of America at the critical time in our narrative—the 1920s.

For Willis, especially in his later papers (e.g. Willis, 1929, 1932), the Chamberlins' hypothesis was self-evidently true. Rocks on earth were too stiff to flow and so the positions of continents and oceans must have remained fixed through geological time. Some "diastrophism" was evident, concentrated at continental margins but widespread peneplanations recorded by regional unconformities were taken as indicative that these deformations were of minor significance. Where there were fossil assemblages that could be correlated between separated continents, both Willis (1928, 1929) and Schuchert (1928) invoked transient land bridges that could rise and fall within otherwise deep-water ocean basins.

That continents and oceans were geographically fixed was not entirely the unanimous view of North American earth scientists. Mobilism was seriously advocated by Reginald Daly (1871–1957) and Frank Bursley Taylor (1860–1938). In Europe however, mobilism was more broadly accepted, with notable Alpine geologists such as Eduard Suess (1831–1914) and Emile Argand (1879–1940) convinced that mountain ranges were produced by substantial lateral motion. Wegener joined the mobilist community, and although his earlier publications, in German, were known to North

American geologists, the tipping point came in 1924, with publication of the English translation *'The origin of continents and oceans'*. Amongst a plethora of evidence and speculations, His pre-eminent case for mobilism came from his correlation of significant geological features from Africa and Europe with those in the Americas. Famously, he argued that the Atlantic Ocean basin was once closed, had opened only in relatively recent geological history and that landmasses had once formed a single supercontinent—Pangea. While the palaeogeographic evidence for continental drift was one just one component for Wegener's thesis—as for other mobilists, the challenge was to find plausible driving mechanisms for tectonics.

Aware of the disagreement, US-based, Dutch petroleum geologist, and a founding member of the American Association of Petroleum Geologists, Willem A. J. M. van Waterschoot van der Gracht (1873–1943) set up a symposium for the evening of 15th November 1926. This event has become notorious in the history of science, as the point in time where North American earth scientists turned their backs on research that supported, what we now know as, plate tectonics. However, as Newman (1995) pointed out, this is an overstatement. Debates, recorded in publications, including reviews of Wegener's book, foreshadowed the symposium. A significant milestone was the publication of a memoir published by AAPG, strictly not the proceedings of the symposium but rather a set of papers solicited and edited by van der Gracht. At the time, physical publications had more longevity in reinforcing scientific opinions than did seminars and so these are our chief concern here.

3 The Language of Argument

Most histories and reports of the Continental Drift Debate argue that the rejection of Wegener's hypothesis was because of a lack of a plausible tectonic driving mechanism and a revocation of the tentative explanations he proposed. As Newman (1995) notes, such emphases miss the point. He identified chauvinism in the hostility towards mobilism by the North American fixists—contrasting the high self-regard of their own intellectually rigorous methods with, in their minds, the loose thinking of Europeans. It was recognised at the time that the debate of continental drift theory revolved around the communities to which the geologists belonged. Van der Gracht's introduction attributes this quote to Reginald Daly *"In Europe its discussion is lively and many geologists as well as geophysicists, are convinced that the hypothesis of continental displacement on a large scale should not be summarily rejected. As yet American geologists have not done their share in developing the possibilities and probabilities of the case"* (Van der Gracht, 1928, p. 42).

Willis (1928) presented a step-wise examination of some of the constituent parts of Wegener's hypothesis, producing counter-arguments. An example is his dismissal of the fit of African and South American continents: *"Its very perfection renders it impossible that there should have been normal faulting along either coast"* (Willis, 1928, p. 80). He made these statements without substantial justification beyond bald expectations. Willis concluded that: *"the theory is presented we find: that the author offers no direct proof of its verity; that the indirect proofs assembled from geology, paleontology, and geophysics prove nothing in regard to drift unless the original postulate of drifting continents be true; that the fields of related sciences have been searched for arguments that would lend color to the adopted theory, whereas facts and principles opposed to it have been ignored. Thus the book leaves the impression that it has been written by an advocate rather than by an impartial investigator"* (Willis, 1928, p. 82). The same criticisms can be made of Willis's own intervention (Newman, 1995).

Over 41 pages, Charles Schuchert (1928) presented a systematic yet highly partial critique of Wegener's hypotheses, his evidence and methods—dismissed as showing *"the nimbleness and versatility of his mind* […] *how very easy it is for him to make all facts fit his hypothesis"* (Charles Schuchert, 1928, p. 134). Wegener is accused of false arguments *"because he generalizes from the generalizations of others* […]*"*. He also selectively quoted the review of Wegener's work by Pierre Termier (1859–1930; director of the Geological Cartography Service of France), that mobilism was *"a beautiful dream, the dream of a great poet. One tries to embrace it, and finds that he has in his arms but a little vapor or smoke; it is at the same time both alluring and intangible"* (Charles Schuchert, 1928, p. 140). Schuchert concluded: *"The battle over the theory of the permanency of the earth's greater features* […] *has been fought and won by Americans long ago. In Europe, however, this battle is not yet fought to a conclusion, since there are leading geologists who* […] *believe in the impermanence of the continents and oceans, and others who do not hesitate to push the earth's poles anywhere in order to explain single floral or faunal peculiarities"* (Charles Schuchert, 1928, pp. 140–141).

Chester Ray Longwell (1887–1975) was, at the time of the debates, a junior colleague of Schuchert's at Yale. He was more measured in appraisal of Wegener, though still noted: *"Certain demands are made of this new and romantic speculation before it is admitted into the respectable circle of geological theories. It must meet the test of established scientific principles, and it must not create more problems than it pretends to solve"* (Longwell, 1928, p. 146).

The most fervent treatment of Wegener's work was provided by Rollin Chamberlin (1928). He accused Wegener of *"dogmatism"*, saying that *"a great deal of Wegener's argumentation seems very superficial"* (Chamberlin, 1928, p. 87), continuing to make a series of (numbered) unsubstantiated assertions. Chamberlin's language was provocative and intemperate, such as *"matching of moraines is ludicrous"*. Of his scientific approach—*"it plays a game in which there are few restrictive rules and no sharply drawn code of conduct"*. And he hid behind unattributed sources in quoting parts of conversations apparently overheard at scientific gatherings: *"Can we call geology a science when there exists such difference of opinion on fundamental matters as to make it possible for such a theory as this to run wild."* And *"If we are to believe Wegener's hypothesis we must forget everything which has been learned in the last 70 years and start all over again"*.

But, as Newman (1995) notes, hostility ran far deeper—unpublished correspondence makes even starker reading. As reported by Oreskes (1999, p. 202), Schuchert wrote to the Gustaaf Molengraaff, Professor at Delft University in the Netherlands:

"to admit that North America and South America have drifted to the West by up to 5000 miles can only be acceptable to forgetful or deranged minds".

Schuchert and Willis coordinated their efforts to promote fixist ideology including papers, published consecutively, in the Bulletin of the Geological Society of America (Schuchert, 1932; Willis, 1932). They even circulated jointly bound offprints. Both Oreskes (1999, p. 218) and Newman (1995, pp. 77–78) report the correspondence from Schuchert to Willis in which they arranged their papers: *"I hope the geologists of the World will agree to a man to accept our views as to how the transoceanic bridges may be made and unmade, not the unreasonable ones asked for by biogeographers* [i.e. mobilists] *crossing all parts of the oceans, but the reasonable ones asked for by palaeontologists! Ha! Ha! Ha! We live or die together, nicht wahr?"* The use of German in concluding the letter was surely not coincidental.

The dogmatism evident in the more hostile critics of Wegener is remarkable given the miniscule amount of hard evidence that was available to earth scientists in the 1920s. The study of whole-earth geophysics was in its infancy. Very little was known of the geology of the ocean floor and even bathymetric maps were rudimentary. Yet Schuchert, Willis and the Chamberlins put theory ahead of empiricism, attacking Wegener's basic observations and first-order geological interpretations while invoking mysterious transient land-bridges to explain faunal distributions. As Newman (1995, p. 79) comments, this was all the more unfortunate as these geologists *"preached open mindedness and multiple working hypotheses"*.

American intransigence is widely regarded as a missed opportunity in the development of modern understanding of global tectonics. Consider the contribution by Molengraaff to the AAPG Memoir: *"the mid-Atlantic ridge is nothing but the* [...] *fracture, along which the disruption of the American continent from the European-African continent took place"* and *"[...] the mid-Atlantic fracture is strictly comparable to the great rift-valley in East Africa"* (Molengraaf, 1928, p. 91). In the English-speaking world, away from North America, the support and continued development of mobilist ideas was restricted to a few geologists such as Arthur Holmes in the UK and Alexander du Toit in South Africa. But overall its denigration remained unchecked, as evidenced by George Lees' presidential address to the Geological Society of London in 1953: *"... the supposed fit of the two sides of the Atlantic can only be achieved by bending and manipulating the continents to such a degree that the whole conception loses validity, unless a jigsaw-puzzle type of manoeuvre is allowed as a scientific method"* (Lees, 1953, p. 234).

Of course, in the 1960s plate tectonics established the validity of much of Wegener's evidence, if not his mechanism, for continental drift. Much of the direct evidence came from the oceans, confirming Molengraaff's proposal of the mid-Atlantic ridge as the locus of rifting (sea floor spreading). Ironically, key developments in improving bathymetric maps were driven by the US Navy, notably by Harry Hess who was educated at Yale by Schuchert and joined the US Navy during WW2 (see for example Hess & Maxwell, 1953). However, palaeomagnetic evidence for continental drift was contested by many in North America as it was developed in the late 1950s and early 1960s (e.g. Vine & Matthews, 1963). Continental geology, and the earth science community in North America finally caught up in the 1960s with the work of those such as Tuzo Wilson (e.g. Wilson, 1966, 1968).

4   Discussion: Lessons from History

The Continental Drift Debate continues to resonate in studies of scientific arguments. It has been looked at as a study in the role of "styles of thought" (Pellegrini, 2019, 2022; Weber & Šešelja, 2020) and how scepticism allied to ridicule were deployed by the fixists to reject mobilist ideas. Allen and Reedy (2020) consider this to be an example of *groupthink*. Defined by Janis (1971) as *"a psychological drive for consensus at any cost that suppresses dissent and appraisal of alternatives in cohesive decision making groups"*, early identifications of groupthink include the analysis of decision-making in US foreign policy (Janis, 1972). It has since been identified as contributing to poor decision-making in some financial institutions (Bénabou, 2013). More recently, groupthink has been evaluated in contributing to diverse disasters, from mountaineering expeditions (Burnette et al., 2011) to the Fukushima accident (Matsui, 2017). And in applied earth science, without using the term, Slater (2023) suggests that fixed thinking bred complacency in risk assessments and associated operational decision-making led to the Deepwater Horizon (Macondo Well) disaster in 2010.

The drift debate has been used as an example of groupthink by Mackey (2023) in his challenge to conventional views on the climate crisis. Hamilton (2019) warns of the dangers of groupthink in perpetuating what he calls "myths" in earth science. He argues that renegade contributions are discouraged. Shanmugam (2022) uses the term groupthink to admonish the broad swathe of sedimentologists publishing on deepwater deposits and processes. We need to take care when drawing lessons from history not to fall into the same behavioural traps that ensnared the fixists. The challenge is that using the term accusatorially may simply reinforce entrenched opinions.

Our contention is that North American antagonism to the descriptions of a tectonically mobile earth is a prime example of groupthink, and that it was explicitly reinforced by the language and tone of the debate. "Othering"—creating polar opposites and forcing a choice between them—is likely to devalue diverse thinking, or certainly its acceptance when weighing choices. We are aware of danger here while characterising the Continental Drift Debate in terms of a conflict between fixists and mobilists. In this we are not alone. Pellegrini (2019) goes further and equates the fixists with religious conservatism and support for biblically derived versions of earth history. Such corollaries may not be helpful. The language used in debate is important—on all sides of any particular argument. Polemics erodes trust in science (König & Jucks, 2019).

The impact of personal ties and personality traits in creating environments within which groupthink can flourish is explored by Riccobono et al. (2016). This certainly applies to our case study. Fernandez (2007) argues that the propensity for groupthink is reduced by increasing the diversity of individuals involved. Too often diversity is seen only from the perspective of the individual. However, the importance of fostering diversity in the composition of financiers striking the balance between risk and reward around investments was characterised as the Lehman Sisters Hypothesis (Van Staveren, 2014). Diversity is expressed simply in terms of the gender balance in the boardroom, but applies equally to other under-represented groups.

As Smith-Doerr et al. (2017) note, broader diversity makes for enhancements in the effectiveness of teams, provided the skills and experiences of a diversity of team members are integrated effectively. So the challenge remains in giving appropriate recognition to innovators in teams that act against groupthink. Hofstra et al. (2020) find that innovators from traditionally under-represented groups (e.g. defined by race or gender) are generally poorly rewarded in comparison with the over-represented groups of scholars. Dissenting voices need to be empowered and certainly not intimidated from speaking out—indeed researchers should strive to challenge orthodoxy. The authors acknowledge that in writing this piece they are leveraging the privileges afforded to them of the academic system under discussion. As white, western people this system has benefited us. We argue for diversification of ideas and in doing so we argue for diversification in peoples. Different genders, nationalities, identities, cultures and backgrounds are required to create true diversity in ideas and minimise groupthink. We challenge the system that feeds us and to be able to do so without fear or retribution is an immense privilege. We recognise that we do not have the answers or solutions and to develop these properly requires collaboration, openness and iteration as a greater diversity in voices and ideas are engaged in earth science.

While there are relatively few discussions of the role of groupthink in the decision process associated with applied geology (but see Phillips, 2011), the importance of assessing uncertainties in the knowledge base for developing a full range of scenarios for risk assessment is noted by Tosoni et al. (2018). Increasingly the subsurface is set to be used to provide sites for the disposal of waste products from power generation—specifically carbon dioxide and radioactive waste. Doing this safely and at scale represent arguably the greatest technical challenges that have faced engineers and earth scientists, both important members of these technical teams. Characterising the subsurface to facilitate engineering projects has been at the forefront of the oil and gas sector and there is an expectation that geoscience expertise will be sequestered from these traditional extractive applications, especially to support CO_2 storage projects. Yet the requirements of new engineering are different—for example as discussed by Jenkins et al. (2024)—so it will be especially important to create diverse teams to manage risk. Those making and implementing policy and the research that underpins this, in earth science and at large, could do worse than reflect on the lessons from the conduct of the Continental Drift Debate.

References

Allen, D. M., & Reedy, E. A. (2020). Seven cases: Examples of how important ideas were initially attacked or ridiculed by the professions. In D.M. Allen, J.W. Howell (Eds.), *Groupthink in science: Greed, pathological altruism, ideology, competition, and culture* (pp. 49–62). Springer. https://doi.org/10.1007/978-3-030-36822-7_5

Bénabou, R. (2013). Groupthink: Collective delusions in organizations and markets. *The Review of Economic Studies, 80*(2), 429–462. Oxford University Press.

Burnette, J. L., Pollack, J. M., & Forsyth, D. R. (2011). Leadership in extreme contexts: A groupthink analysis of the May 1996 Mount Everest disaster. *Journal of Leadership Studies, 4*(4), 29–40. https://doi.org/10.1002/jls.20190

Chamberlin, R. T. (1928). Some of the objections to Wegener's theory. In V. AJv. W. van der Gracht (Ed.), *Theory of continental drift: A symposium* (pp. 83–87). University of Chicago Press.

Fernandez, C. P. (2007). Creating thought diversity: The antidote to group think. *Journal of Public Health Management and Practice, 13*(6), 670–671. https://doi.org/10.1097/01.PHH.000029 6146.09918.30

Frankel, H. R. (1985). The biogeographical aspect of the debate over continental drift. *Earth Sciences History, 4*(2), 160–181. https://doi.org/10.17704/eshi.4.2.t05v18730x38g462

Frankel, H. R. (2012). *The continental drift controversy (4 volumes)*. Cambridge University Press.

Hamilton, W. B. (2019). Toward a myth-free geodynamic history of earth and its neighbors. *Earth-Science Reviews, 198*, Article 102905. https://doi.org/10.1016/j.earscirev.2019.102905

Hess, H. H., & Maxwell, J. C. (1953). Major structural features of the south-west Pacific: A preliminary interpretation of H.O. 5484, bathymetric chart, New Guinea to New Zealand. In *Proceedings of the 7th Pacific Science Congress*, 14–17.

Hofstra, B., Kulkarni, V. V., Munoz-Najar Galvez, S., He, B., Jurafsky, D., et al. (2020). The diversity–innovation paradox in science. *PNAS, 117*(17), 9284–9291. https://doi.org/10.1073/pnas.1915378117

Janis, I. L. (1971). Groupthink. *Psychology Today* 84–90. Retrieved June 10, 2025, from https://agcommtheory.pbworks.com/f/GroupThink.pdf

Janis, I. L. (1972). *Victims of groupthink; A psychological study of foreign-policy decisions and fiascoes*. Houghton, Mifflin.

Jenkins, C., Pestman, P., Carragher, P., Constable, R. (2024). Long-term risk assessment of subsurface carbon storage: Analogues, workflows and quantification. *Geoenergy, 2*. https://doi.org/10.1144/geoenergy2024-014

König, L., & Jucks, R. (2019). Hot topics in science communication: Aggressive language decreases trustworthiness and credibility in scientific debates. *Public Understanding of Science, 28*(4), 401–416. https://doi.org/10.1177/0963662519833903

Le Grand, H. E. (1986). Steady as a rock: Methodology and moving continents. In J. Schuster & R. R. Yeo (Eds.), *The politics and rhetoric of scientific method: Historical studies* (pp. 97–138). Springer, Netherlands.

Lees, G. M. (1953). The evolution of a shrinking earth. *Quarterly Journal of the Geological Society, 109*(1–4), 217–257. https://doi.org/10.1144/GSL.JGS.1953.109.01-04.10

Longwell, C. R. (1928). Some physical tests of the displacement hypothesis. In V.A.J. vW. van der Gracht (Ed.), *Theory of continental drift: A symposium* (pp. 145–157). University of Chicago Press.

Mackey, R. (2023). Lessons from the continental drift controversy for the IPCC debacle. *Science of Climate Change, 3*, 487–520. Retrieved June 10, 2025, from https://scienceofclimatechange.org/wp-content/uploads/Mackey-2023-Continental-Drift-IPCC.pdf

Matsui, R. (2017). Groupthink trap: A study on the essence of failure in Fukushima nuclear accident. *Transactions of the Academic Association for Organizational Science, 6*(2), 14–19. https://doi.org/10.11207/taaos.6.2_14

Molengraaf. (1928). Wegener's continental drift. In V.A.J. vW. van der Gracht (Ed.), *Theory of continental drift: A symposium* (pp. 90–92). University of Chicago Press.

Newman, R. P. (1995). American intransigence: The rejection of continental drift in the great debates of the 1920s. *Earth Sciences History, 14*(1), 62–83. https://doi.org/10.17704/eshi.14.1.866486 2716123568

Oreskes, N. (1999). *The rejection of continental drift: Theory and method in American earth science* (p. 420). Oxford University Press.

Pellegrini, P. A. (2019). Styles of thought on the continental drift debate. *Journal for General Philosophy of Science, 50*, 85–102. https://doi.org/10.1007/s10838-018-9439-7

Pellegrini, P. A. (2022). About the reaction to styles of thought on the continental drift debate. *Journal for General Philosophy of Science, 53*, 573–582. https://doi.org/10.1007/s10838-022-09617-2

Pettijohn, F. J. (1970). Rollin Thomas Chamberlin 1881–1948: A Biographical Memoir. National Academy of Sciences, Washington, D.C. Biographical Memoirs. Retrieved June 10, 2025, from https://www.nasonline.org/wp-content/uploads/2024/06/chamberlin-rollin-t.pdf

Phillips, G. N. (2011). Gold exploration success. *Applied Earth Science, 120*(1), 7–20. https://doi.org/10.1179/1743275811Y.0000000019

Riccobono, F., Bruccoleri, M., & Größler, A. (2016). Groupthink and project performance: The influence of personal traits and interpersonal ties. *Production and Operations Management, 25*(4), 609–629. https://doi.org/10.1111/poms.12431

Romano, M., Console, F., Pantaloni, M., & Fröbisch, J. (2017). One hundred years of continental drift: The early Italian reaction to Wegener's 'visionary' theory. *Historical Biology, 29*(2), 266–287. https://doi.org/10.1080/08912963.2016.1156677

Schuchert, C. (1928). The hypothesis of continental displacement. In V.A.J. vW. van der Gracht (Ed.), *Theory of continental drift: A symposium* (pp. 104–144). University of Chicago Press.

Schuchert, C. (1932). Gondwana land bridges. *GSA Bulletin, 43*(4), 875–916. https://doi.org/10.1130/GSAB-43-875

Shanmugam, G. (2022). 150 Years (1872–2022) of research on deep–water processes, deposits, settings, triggers, and deformation: A difficult domain of progress, dichotomy, diversion, omission, and groupthink. *Journal of Palaeogeography, 11*(4), 469–564. https://doi.org/10.1016/j.jop.2022.08.004

Slater, D. H. (2023). Was the deepwater horizon incident a "Normal" accident? *Safety Science, 168*, Article 106290. https://doi.org/10.1016/j.ssci.2023.106290

Smith-Doerr, L., Alegria, S. N., & Sacco, T. (2017). How diversity matters in the US science and engineering workforce: A critical review considering integration in teams, fields, and organizational contexts. *Engaging Science, Technology, and Society, 3*, 139–153. https://doi.org/10.17351/ests2017.142

Tosoni, E., Salo, A., & Zio, E. (2018). Scenario analysis for the safety assessment of nuclear waste repositories: A critical review. *Risk Analysis, 38*(4), 755–776. https://doi.org/10.1111/risa.12889

van der Gracht, V. A. J. vW. (1928). The problem of continental drift. In V.A.J. vW. van der Gracht (Ed) *Theory of continental drift: A symposium* (pp. 1–75). University of Chicago Press.

Van Staveren, I. (2014). The Lehman Sisters hypothesis. *Cambridge Journal of Economics, 38*(5), 995–1014. https://doi.org/10.1093/cje/beu010

Vine, F. J., & Matthews, D. H. (1963). Magnetic anomalies over oceanic ridges. *Nature, 199*, 947–949. https://doi.org/10.1038/199947a0

Weber, E., & Šešelja, D. (2020). In defence of rationalist accounts of the continental drift debate: A response to Pellegrini. *Journal for General Philosophy of Science, 51*, 481–490. https://doi.org/10.1007/s10838-020-09516-4

Wegener, A. (1924). *The origin of continents and oceans* (translated by Skerl, J.G.A.). Methuen and Co., London.

Willis, B. (1928). Continental drift. In V.A.J. vW. van der Gracht (Ed.), *Theory of continental drift: A symposium* (pp. 76–82). University of Chicago Press.

Willis, B. (1929). Continental genesis. *GSA Bulletin, 40*(1), 281–336. https://doi.org/10.1130/GSAB-40-281

Willis, B. (1932). Isthmian links. *GSA Bulletin, 43*(4), 917–952. https://doi.org/10.1130/GSAB-43-917

Wilson, J. T. (1966). Did the Atlantic close and then re-open? *Nature, 211*, 676–681. https://doi.org/10.1038/211676a0

Wilson, J. T. (1968). Static or mobile earth: The current scientific revolution. *Proceedings of the American Philosophical Society, 112*(5), 309–320.

Towards Integrating Indigeneity and Geoethical Practice in Aotearoa New Zealand

Matthew W. Hughes⑩ **and Norm Kereikeepa**

Abstract The place-based knowledges of Indigenous communities can provide rich contributions to geoethical thought and practice. Traditional Indigenous Māori beliefs in Aotearoa New Zealand are animist and holistic and assume the existence of a meta-physical realm that imbues landscape features with a spiritual life essence; these beliefs inform a deep guardianship and sustainability ethic, and the agency ascribed to landscape features provides an Indigenous sustainability framework. The pluralist nature of geoethics means that geoscientists and engineers should be able to respect-fully acknowledge Māori beliefs and connection to landscape to support meaningful community engagement, to mitigate negative impacts of ground disturbance from civil engineering works and to inform infrastructure design and construction. We present a case study of post-earthquake infrastructure reinstatement to explore these concepts and show that legislative and infrastructure recovery responses to disasters can be more harmful to Indigenous landscapes, comprised of sacred landforms and sites, than the natural event itself. In these contexts, engineers and other actors have (geo)ethical obligations that must balance wider societal needs and Indigenous rights and interests.

Keywords Aotearoa New Zealand · Indigenous · Māori · Landscape · Infrastructure

1 Introduction

Geoethics, the field of theoretical and applied ethics focused on the human–Earth system nexus, is universal (i.e. for wider humanity) and has global and regional/local dimensions (Di Capua & Peppoloni, 2023). While universalist and global approaches

M. W. Hughes (✉)
Department of Civil and Environmental Engineering, University of Canterbury, Christchurch, New Zealand
e-mail: matthew.hughes@canterbury.ac.nz

N. Kereikeepa
Te Rūnanga o Kaikōura, Kaikōura, New Zealand

address human species- and planetary-scale concerns, the localism and place-based knowledges inherent in Indigenous communities, gained over ecological timeframes throughout the Late Quaternary and Holocene, can also provide rich contributions to geoethical thought and practice. As articulated by the United Nations Permanent Forum on Indigenous Issues, Indigenous peoples have strong links to their territories and natural resources and will maintain and reproduce their ancestral environments and systems as distinctive peoples and communities. The United Nations Declaration on the Rights of Indigenous Peoples states they also have the right to maintain and strengthen their distinctive spiritual relationships with traditionally owned or otherwise occupied and used lands and waters, and to uphold their responsibilities to future generations in this regard. These concerns and rights of Indigenous peoples are therefore of geoethical interest pertaining to natural resource use, sustainable economic and community development and connections to landscape.

In a recent critique, Paden (2024) responded to Peppoloni and Di Capua's (2023) arguments that non-living geological entities have intrinsic moral importance and are worthy of "moral considerability". Paden (2024) states that these arguments are teleological moral theories addressing goals and purposes of things. Something morally considerable must possess its own welfare driven by goals or purpose, and this welfare must be considered when deliberating about actions affecting it. Conversely, entities without agency or purpose have no welfare to protect or advance and therefore they are not morally considerable and have instrumental value only for the purposes of morally considerable beings (Paden, 2024). From these ideas arise various ethical positions: anthropocentric ethics in which only individual humans have moral importance; biocentric ethics in which sentient nonhuman animals and plants (with purposes such as growth and reproduction) have moral importance; and ecocentric ethics in which ecosystems have unconscious purposes such as supporting biodiversity and climate regulation. However, Paden (2024) states this teleological approach is inapplicable to rocks and geological formations as they are inanimate and non-sentient and lack goals and purpose, and concludes that teleology is unsuitable as a philosophical foundation for geocentric ethics. Instead, this foundation must be sought in virtue ethics.

Within the traditional worldviews of many Indigenous peoples, teleological thinking is common and supports deep ties to landscapes. As stated by Di Capua and Peppoloni (2023), geoethics is by definition pluralist and respects a diversity of visions, approaches and tools to ensure human dignity and guarantee opportunities for effective actions to face common threats; therefore, while geoscientists and engineers themselves need not adhere to these teleological views, an appreciation of them is fundamental to geoethical practice. Indigenous peoples globally have often experienced the harmful effects of infrastructure development, particularly in the mineral resources sector, without receiving benefits from these activities (e.g. see Fernholz, 2010; Horowitz et al., 2024), and as such developments depend on the geosciences and engineering, these professions bear particular geoethical responsibilities. These ideas are explored below in the context of Aotearoa New Zealand. First, we present an overview of traditional animistic and holistic worldviews of Māori, the Indigenous peoples of Aotearoa New Zealand, pertaining to landscapes and geological and geomorphological features. We then present a case study of conflict between Māori landscape values and post-disaster infrastructure recovery affecting Indigenous landscapes. We conclude with pathways for improved recognition of Indigenous landscape values in such contexts.

2   Animism and Holism in the Traditional Māori Worldview

Prior to the arrival of Christianity to Aotearoa New Zealand through the British colonial project, with its specific conceptions of space, time and relationships with the natural world, Māori had their own distinct philosophies brought by their Polynesian ancestors and subsequently informed by the characteristics of their new land (Salmond, 2017). Fundamental were "nature narratives" of creation and origins that served as theories about reality, and as philosophies giving rise to ethics and values (Stewart, 2020, p. 60). Individual entities could only be conceived of with respect to their relations with other entities, an epistemological approach termed *animism* and founded on the understanding that *"relationships humans have with nonhuman entities are reciprocal and contextual rather than unidirectional and abstract, and that as these relationships progress each entity shapes the other in meaningful ways"* (Reid & Rout, 2016, emphasis in original). This reflects an Indigenous philosophical approach that looks *"most essentially at how one is positioned in relation to another"*, and that other could be a person, a mountain or a tree (Mika, 2015). Further, as stated by Mika (2012): *"Māori believe that the self is part of the environment, and hence the self's uptake of anything – emotion, feeling, cognition, even physical attribute – is dependent on the interplay of whakapapa* ["genealogy"; see below] *with the natural world. The deep links that Māori have with the natural world – seen and unseen – permeate outwards to include those who are deceased and those who are yet to come, as well as past and future impacts on the environment"*. Animism's view of reality is one comprised of a complex and dynamic mixture of interacting subject–object hybrids needing constant contextual interpretation (Rout & Reid, 2020). In the Māori worldview, this leads to a sustainability ethic comprised of intrinsic and instrumental motivations: intrinsic because every (nonhuman) entity that is interacted with is deserving of respectful relations; and instrumental because mutual interdependence implies that human impacts on ecosystems in turn impact human wellbeing over multiple timeframes, motivating long-term approaches to sustainable action (Rout & Reid, 2020).

The holistic, interconnected nature of reality in the Māori worldview is exemplified in these key concepts: *whakapapa, tīpuna, mauri, wairua, kaitiakitanga* and *whenua. Whakapapa* can be translated as "genealogy", and to Māori everything is connected genealogically; humans to each other, humans to other life forms and humans to soils, rocks and landscape features. In traditional Māori ordering of nature, genealogies were developed to explain how different types of rocks and soils relate to each other, and how they were manifestations of deities and their progeny. In Māori mythology, the deity Tane, son of Papatūānuku the Earth Mother, fashioned the first women from clay, and she was named Hineahuone, meaning earth-formed woman; therefore, in this Māori creation tradition, all humans can be traced genealogically to the substance of the Earth itself. *Tīpuna* is the Māori concept of "ancestors", those who gave rise to current generations, including phenomena that supported the ancestors' existence such as rivers (Reid & Rout, 2016). In the traditional Māori worldview, humans descend from both their human forebears and the ecosystems, soils and rocks comprising the landscapes on which they reside, connected through *whakapapa* (genealogy). *Mauri* can be translated as "life essence" and refers to the health and vitality of a particular entity; all beings and environmental phenomena are believed to be animated by *mauri* (Reid & Rout, 2016). *Mauri* can be enhanced or degraded, such as the health of rivers and soils affected positively or negatively by external forces, and is also interdependent in that an entity's life essence, if dependent on that of another, can be impacted by the status of the supporting entity's *mauri*. *Mauri* animates animism, guides and shapes relations among all beings and entities (Rout & Reid, 2020), and a Māori environmental ethic builds relationships that enhance *mauri* (Reid & Rout, 2016). A related concept is *wairua* or non-physical "spirit" distinct from the human body, and some believe that inanimate things also possess *wairua*. *Kaitiakitanga* refers to "guardianship" and the responsibility to safeguard environmental integrity and protect and foster *mauri*. This guardianship emphasises humans being integral and dependent components of the environment. In modern parlance, *kaitiakitanga* can be considered sustainable environmental management based on a holistic view of living ecological systems (Reid & Rout, 2016). Finally, *whenua* refers to "land" and is also the Māori word for "placenta", thus expressing the inextricable relationship between people and land, cemented through burial of the placenta in the place of one's birth. A key descriptor of Māori kin groups within a particular geographic context is *tangata whenua*—people of the land.

The references above to worlds unseen, relationships with deceased ancestors and unborn descendants, the *mauri* "life essence" and *wairua* "spirit", reflect an explicit acceptance in the traditional Māori worldview of a metaphysical realm permeating the perceptible physical realm. As stated by Mika (2015), within much Indigenous philosophy metaphysics *"must be thought of as an integral, infusing element within the world"*. Such thinking contrasts with rationalist scientific objectivity, in which metaphysical explanations of reality are not considered explanations at all. In a critique of the concept of *mātauranga*, a non-traditional Māori term equated with "knowledge", Mika (2012) states that *"Māori believe there is a sense of mystery to the world"*, and posits that a focus on such a knowledge system *"fixes things in the world that Māori cite are meant to be mysterious to the self"* and *"is a result of epistemic certainty that detracts from a Māori focus on Being* [oneness of existence]", which as *"a traditional phenomenon…emphasises mystery and unknowability"*. This ontological position also conflicts with scientific approaches grounded in the potential knowability of material reality and that reject the notion that phenomena are "meant" to be mysterious.

To summarise, in traditional Māori animist thought, Earth systems—inseparable from an underpinning metaphysical realm—can be ancestors imbued with agency; this provides an Indigenous teleological framework for perceiving and relating to these systems. These conceptual approaches appear to challenge scientific modes of thinking that attempt to remove mystery and reduce uncertainty. However, Indigenous views on the unity of the physical world have echoes in modern Earth, ecosystem and sustainability sciences, and these holistic parallels should not be dismissed because of belief in metaphysical foundations. Geoethicists need not adhere to these ontological and epistemological views but should account for them when conducting works directly or indirectly affecting Indigenous communities. Geoethics recognises the plurality of human belief and knowledge systems and aims to synthesise different existential conceptions of relationships between human beings and the Earth system (Di Capua & Peppoloni, 2023). Virtue ethics depends on the education and positive character development of geoethicists able to make complex judgements to support morally appropriate actions (Paden, 2024), and this can include respectful recognition of different worldviews, including animist ones. In Aotearoa New Zealand, for geomorphological research and exploring Indigenous relationships with dynamic landscapes, attempts have begun to better integrate science and Māori worldviews (Wilkinson et al., 2020, 2021). Yet, in the context of civil engineering projects that modify landscapes significant to Māori, there is still much scope for development and operationalisation of geoethical principles.

3 Post-disaster Infrastructure Recovery in Indigenous Landscapes

Centred on the Kaikōura coastal region of Aotearoa New Zealand is the ancestral homeland of Ngāti Kuri, a *hapū* (subtribe; extended kin network) within the wider Ngāi Tahu *iwi* (tribe). Centuries of continuous use and occupation have conferred to Ngāti Kuri customary/ancestral authority over their lands and seas, and on the coastal margins and further inland created a rich Indigenous landscape of *wāhi tapu* (sacred sites), terraformed hilltops formerly the sites of *pā* (fortified villages) and remnants of occupation sites and marine resource processing evidenced in middens. The area contains traditional interment sites, and along the open rocky coast and within sheltered bays are burial grounds holding those killed in historical conflicts. Because seafaring for fishing and voyaging along the coast was so important, rocky headland landforms were critical waypoints for navigation; sacred "talisman rocks" signifying safe harbour and proximity to home. In terms described in the previous section, Ngāti Kuri have deep *whakapapa* (genealogical connection) with the area in which their human *tīpuna* (ancestors) are interred. Because of their importance for safe navigation, the coastal talisman rocks are *wāhi tapu* imbued with *mauri* (life essence) and *wairua* (spirit). While many sites are documented in modern archaeological databases, this information is a significant under-representation of sites known to the Ngāti Kuri *hapū*.

Ngāti Kuri's *kaitiakitanga* (guardianship) values and policies are expressed in their Environmental Management Plan (Te Rūnanga o Kaikōura, 2007), which documents *hapū* approaches to safeguarding terrestrial and marine environments, and to engaging with other societal actors. The Environmental Management Plan is an *iwi* (tribal) planning document, and under the Resource Management Act (RMA) 1991 is applicable to planning processes undertaken by local government. Of relevance here are Ngāti Kuri's concerns for earthworks, including protection of *wāhi tapu*, archaeological sites and landscape values from disturbance, cultural monitoring of earthworks and appropriate processes for accidental discovery of cultural materials. Modification, damage or destruction of archaeological sites in the area must not be approved without consultation and approval from Te Rūnanga o Kaikōura (the administrative council of Ngāti Kuri). Further, Ngāti Kuri expertise, knowledge and oral traditions relating to cultural heritage must be recognised, respected and provided for by local government, and inclusion of *hapū* experts may be required where excavations pose risks to archaeological and heritage sites.

On 14th November 2016, the moment magnitude 7.8 Hurunui/Kaikōura Earthquake struck northeastern regions of the South Island of Aotearoa New Zealand (Bradley et al., 2017). The series of causative fault ruptures and seismic directivity resulted in a concentration of landscape and infrastructure impacts in Kaikōura's mountainous coastal areas. Fault ruptures caused large vertical and horizontal ground displacements (Hamling et al., 2017), and significant ground shaking led to thousands of shallow and deep-seated landslides (Dellow et al., 2017). The fault ruptures and landslides caused extensive damage to road and railway infrastructure systems comprising the coastal sections of State Highway 1 and the Main North Line railway (Davies et al., 2017); both systems were the main freight and tourist routes for this part of the country. Kaikōura township, an agricultural service hub and popular tourist destination on the coastal route, was isolated with significant negative impacts on these sectors and the psycho-social wellbeing of local communities (Fountain & Cradock-Henry, 2023).

On 12th December 2016, the Aotearoa New Zealand central government issued a legislative response to the earthquake—the *Hurunui/Kaikōura Earthquakes Recovery Act*, which provided for introduction of the *Hurunui/Kaikōura Earthquakes Recovery (Coastal Route and Other Matters) Order 2016*. This was what is termed an Order in Council, designed to enable government ministries to rapidly amend legislation they are responsible for. The Order in Council significantly amended the RMA 1991, which, for large-scale civil infrastructure works in non-disaster contexts, would require environmental impact assessments to be conducted, invitation of public submissions on proposed works and requirements for community engagement in specific contexts. The RMA 1991 also has specific regard for Māori interests and, as mentioned above, recognises Māori environmental management plans. The Order in Council was implemented to expedite recovery of coastal transportation infrastructure deemed of national significance and to re-establish connectivity as soon as possible, but it removed legislative compulsion for detailed environmental impact assessments and community engagement in general, and negated *kaitiakitanga* policies of Ngāti Kuri in particular.

The North Canterbury Transport Infrastructure Recovery (NCTIR) organisation was established in December 2016 to clear landslide debris, stabilise landforms and reinstate transportation routes, and by March 2018, when the Order in Council was revoked, had completed most of the required geomechanical remediation and civil infrastructure works. However, NCTIR included no Ngāti Kuri representatives at the project planning stages nor made provision for dedicated cultural monitoring, and it was over this time that most negative impacts occurred to archaeological and heritage sites and *wāhi tapu* from ground disturbance. As observed by one of us (Kereikeepa) in his role as Ngāti Kuri cultural monitor, while shaking and landsliding from the Hurunui/Kaikōura Earthquake caused modifications to five sites, over the Order in Council period several sites were destroyed and a large number modified. Disturbances included access track construction through a *pā* complex, destruction of a known archaeological site containing historic gardens, severe disturbance of a documented mass burial site known to district and regional authorities for decades, and the destruction of one coastal talisman rock and modification of three others. The destruction of known archaeological and *wāhi tapu* sites was particularly egregious considering the availability of this information for planning and design of civil infrastructure works under normal RMA 1991 processes. Nevertheless, despite the Order in Council mechanism discouraging community engagement, instances of ethical professionalism did occur in this respect. For a period, NCTIR leadership required infrastructure plans to be shared with the Ngāti Kuri cultural monitor for feedback to mitigate impacts from design and construction, and individual project managers and senior engineers also proactively engaged the cultural monitor. This process was not without flaws; the cultural monitor assented to road reinstatement in particular areas based on the plans presented, but when construction was underway, it became clear that an additional shared pedestrian/bicycle pathway had been included without Ngāti Kuri knowledge or consent—completion of the project would have resulted in more disturbance of coastal burial grounds and the destruction of at least one more talisman rock. Only after revocation of the Order in Council in March 2018 was Ngāti Kuri *hapū* able to intervene via mechanisms of the RMA 1991 and permanently halt the shared pathway construction.

Following the Order in Council revocation, and re-establishment of community engagement and impact assessment processes that recognised the Ngāti Kuri Environmental Management Plan, unnecessary disturbance of heritage/archaeological sites and *wāhi tapu* decreased significantly. In one instance, meaningful and respectful engagement led to redesign of a proposed coastal access area where a burial ground is located, so that the public amenity was preserved while preventing any disturbance of the bodies that still lie there. A visible expression of the respectful engagement between Ngāti Kuri and the infrastructure rebuild project is the Cultural Artwork Package completed in December 2020, a series of Māori artworks along the Kaikōura coast that includes murals depicting tribal narratives on railway tunnels and highway retaining walls, carved wooden and stone *pou* (statues) depicting chiefly ancestors and identifying significant sites, and information panels describing Ngāti Kuri histories. However, while the Cultural Artwork Package is a laudable and welcome expression of Ngāti Kuri identity as *tangata whenua* (people of the land), its presence cannot assuage the pain felt by the community caused by what they view as desecration of their ancestral landscapes due to initial suspension of the RMA 1991 and subsequent poor engagement.

This story of earthquake and infrastructure recovery impacts on Māori communities raises key themes relevant to geoethical practice in the context of civil engineering works and Indigenous landscapes. First, the Māori animist and metaphysical worldview of landscapes and geological features is a real motivating force in community beliefs and behaviours, and significantly informs a guardianship and sustainability ethic; Māori communities are therefore heavily invested in any human activities that cause disturbance to these Earth systems. Second, while the 2016 Hurunui/Kaikōura Earthquake fits the definition of a disaster in that it was an unpredictable rapid-onset event causing overwhelming disruption requiring significant external resources for response and recovery, in fact for the Ngāti Kuri *hapū* the infrastructure recovery was the real disaster; avoidable destruction and alteration of sacred landforms and other sites of cultural value due to infrastructure works. Third, legislation enacted to expedite infrastructure system reinstatement may support wider socioeconomic recovery, but this can conflict with the rights and interests of local Indigenous communities; legislative frameworks have material impacts and can lead to heavy machinery carelessly pushed through known burial grounds and sacred landforms. Fourth, in contexts where community engagement is not mandated, the individual ethical behaviours of engineers in decision-making roles can still significantly influence the course of infrastructure planning and design to minimise disturbance and destruction of archaeological, heritage and other sacred sites. These engineers need not personally adhere to animist metaphysical beliefs in which landforms have a spiritual life essence, but respectful acknowledgement of these beliefs held by communities may lead to post-disaster re-establishment of infrastructure systems that benefit all members of the community, while minimising harm to cultural heritage and landforms in Indigenous landscapes.

4   Conclusion

The traditional beliefs of Māori communities in Aotearoa New Zealand are animist and holistic and assume the existence of a metaphysical realm that imbues landscape features with a spiritual life essence; these beliefs have informed, and continue to inform, a deep guardianship and sustainability ethic that sees human beings as integral and dependent components of the environment. This animist worldview acknowledges geological and other landscape features as ancestors because of the generative and supporting roles they play in human survival. Geoscientists and civil engineers depend professionally on scientific objectivism seemingly incompatible with metaphysics-based animism, yet by acknowledging sincerely held Māori beliefs to support meaningful community engagement on landscape impacts of civil engineering works, and using the knowledge gained to inform design and construction, geoscientists and civil engineers are practising a pluralistic virtue geoethics. In non-disaster contexts in Aotearoa New Zealand, there is already considerable scope for engaging and operationalising Māori worldviews in infrastructure development, and this has long been mandated under the RMA 1991. However, in post-disaster situations, geoethical issues arise from conflicting priorities, such as discharging professional engineering duties mandated under legislation versus respecting the rights and interests of Indigenous communities. As the post-2016 Hurunui/Kaikōura Earthquake example presented here shows, egregious disturbance and destruction of important sites within Indigenous landscapes stemmed directly from modification of the RMA 1991 with no attendant administrative mechanism to engage with local Māori communities in project planning nor cultural monitoring. Nevertheless, even under legislative restrictions, there is space for engagement depending on the personal and ethical motivations of individuals within the engineering (and geoscience) community.

To build geoethical capacity in the engineering, geoscience and policy professions in Aotearoa New Zealand, focusing on the consequences for Māori communities from both business-as-usual infrastructure development projects and post-disaster infrastructure recovery, we suggest the following pathways. First, geoethical principles should be delivered formally in engineering education curriculums at the undergraduate and postgraduate levels, presenting contrasts and complementarities between scientific objectivism and Indigenous animism, and how these can inform successful community engagement. Second, industry accreditation for professional engineers involved in large-scale civil engineering works resulting in significant landscape modification should require professional development in Māori engagement protocols, animist perceptions of Earth systems and Indigenous archaeological/heritage aspects of landscapes. Third, implementation of appropriate (geo)ethical training for personnel within responsible ministries for post-disaster response and recovery, to anticipate community impacts of legislative changes and communicate those to political decision-makers. Fourth, compiling detailed case studies of infrastructure developments and associated negative and positive impacts for Māori communities, with key lessons learned through a (geo)ethical lens. These and other initiatives will in turn inform development of sustainable and resilient infrastructure that supports the aspirations of Māori communities.

Acknowledgements This work was supported by a Ngā Puanga Pūtaiao Research Fellowship awarded to Hughes by the Royal Society of New Zealand/Te Apārangi. This project was also supported by Te Hiranga Rū QuakeCoRE, an Aotearoa New Zealand Tertiary Education Commission-funded Centre. This is QuakeCoRE publication number 1023. Carl Mika provided valuable feedback on the manuscript.

References

Bradley, B. A., Razafindrakoto, H. N., & Nazer, M. A. (2017). Strong ground motion observations of engineering interest from the 14 November 2016 Mw7.8 Kaikōura, New Zealand earthquake. *Bulletin of the New Zealand Society for Earthquake Engineering, 50*(2), 85–93. https://doi.org/10.5459/bnzsee.50.2.85-93

Davies, A. J., Sadashiva, V., Aghababaei, M., Barnhill, D., Costello, S. B., et al. (2017). Transport infrastructure performance and management in the South Island of New Zealand, during the first 100 days following the 2016 Mw 7.8 "Kaikōura" earthquake. *Bulletin of the New Zealand Society for Earthquake Engineering, 50*(2), 271–299. https://doi.org/10.5459/bnzsee.50.2.271-299

Dellow, S., Massey, C., Cox, S., Archibald, G., Begg, J., et al. (2017). Landslides caused by the Mw7.8 Kaikōura earthquake and the immediate response. *Bulletin of the New Zealand Society for Earthquake Engineering, 50*(2), 106–116. https://doi.org/10.5459/bnzsee.50.2.106-116

Di Capua, G., & Peppoloni, S. (2023). An expanded definition of geoethics. EGU General Assembly 2023, Vienna, Austria, 24–28 April 2023, EGU23-1385. https://doi.org/10.5194/egusphere-egu23-1385

Fernholz, R. M. (2010). Infrastructure and inclusive development through "Free, Prior, and Informed Consent" of Indigenous peoples. In W. Ascher & C. Krupp (Eds.), *Physical infrastructure development: Balancing the growth, equity, and environmental imperatives* (pp. 225–258). Palgrave Macmillan. https://doi.org/10.1057/9780230107670_9

Fountain, J., & Cradock-Henry, N. A. (2023). We're all in this together? Community resilience and recovery in Kaikōura following the 2016 Kaikōura-Hurunui earthquake. *New Zealand Journal of Geology and Geophysics, 66*(2), 62–176. https://doi.org/10.1080/00288306.2023.2167842

Hamling, I. J., Hreinsdóttir, S., Clark, K., Elliott, J., Liang, C., et al. (2017). Complex multifault rupture during the 2016 Kaikōura earthquake, New Zealand. *Science, 356*(6334). https://doi.org/10.1126/science.aam7194

Horowitz, L. S., Keeling, A., Lévesque, F., Rodon, T., Schott, S., et al. (2024). Indigenous peoples' relationships to large-scale mining in post/colonial contexts: Toward multidisciplinary comparative perspectives. In T. Rodon, S. Thériault, A. Keeling, S., Bouard, & A. Taylor (Eds.), *Mining and Indigenous livelihoods: Rights, revenues, and resistance* (pp. 109–136). Routledge. https://doi.org/10.4324/9781003406433-8

Mika, C. (2012). Overcoming 'being' in favour of knowledge: The fixing effect of 'mātauranga.' *Educational Philosophy and Theory, 44*(10), 1080–1092. https://doi.org/10.1111/j.1469-5812.2011.00771.x

Mika, C. (2015). Counter-colonial and philosophical claims: An indigenous observation of Western philosophy. *Educational Philosophy and Theory, 47*(11), 1136–1142. https://doi.org/10.1080/00131857.2014.991498

Paden, R. (2024). Order and place in environmental ethics and esthetics. In S. Peppoloni & G. Di Capua (Eds.), *Geoethics for the future: Facing global challenges* (pp. 15–27). Elsevier. https://doi.org/10.1016/B978-0-443-15654-0.00016-5

Peppoloni, S., & Di Capua, G. (2023). The significance of geotourism through the lens of geoethics. In M. Allan & R. Dowling (Eds.), *Geotourism in the Middle East* (pp. 41–52). Springer. https://doi.org/10.1007/978-3-031-24170-3_3

Reid, J., & Rout, M. (2016). Getting to know your food: The insights of indigenous thinking in food provenance. *Agriculture and Human Values, 35*, 283–294. https://doi.org/10.1007/s10460-015-9617-8

Rout, M., & Reid, J. (2020). Embracing indigenous metaphors: A new/old way of thinking about sustainability. *Sustainability Science, 15*, 945–954. https://doi.org/10.1007/s11625-020-00783-0

Salmond, A. (2017). *Tears of Rangi: Experiments across worlds.* Auckland University Press.

Stewart, G. T. (2020). *Māori Philosophy: Indigenous thinking from Aotearoa.* Bloomsbury Academic.

Te Rūnanga o Kaikōura. (2007). Te Poha o Tohu Raumati, Te Rūnanga o Kaikōura Environmental Management Plan (2nd ed.). Retrieved June 10, 2025, from https://ngaitahu.iwi.nz/assets/Doc uments/Te-Runanga-o-Kaikoura-Environmental-Management-Plan.pdf

Wilkinson, C., Hikuroa, D. C. H., Macfarlane, A. H., & Hughes, M. W. (2020). Mātauranga Māori in geomorphology: Existing frameworks, case studies, and recommendations for incorporating Indigenous knowledge in Earth science. *Earth Surface Dynamics, 8*(3), 595–618. https://doi. org/10.5194/esurf-8-595-2020

Wilkinson, C., Macfarlane, A. H., Hikuroa, D. C. H., McConchie, C., Payne, M., et al. (2021). Landscape change as a platform for environmental and social healing. *Kōtuitui: New Zealand Journal of Social Sciences Online, 17*(3), 352–377. https://doi.org/10.1080/1177083X.2021. 2003826

Operational Ethics for Researching Geohazards

Ilan Kelman⊙

Abstract Researching and publishing on geohazards can be hazardous to researchers and to others, leading to operational ethics concerns about balancing the need for geohazards research and the need for everyone's safety. Potential ethical dangers of geohazards research are often known, identified and accepted, for instance through institutional risk assessment and research ethics processes. Not all possible ethical conundrums are necessarily fully accounted for, especially since the possibilities are physical and social. Identifying these operational ethical issues, in theory and in practice, is typically straightforward. Resolving the concerns tends to be more difficult. Science diplomacy within the context of disaster diplomacy research demonstrates how operational ethics for researching geohazards might be improved, at least in theory. Examples to consider are the usefulness of codes of conduct for improving ethical behaviour, the balance between risk reduction and risk management and the effectiveness of monitoring, enforcement and accountability mechanisms. In practice, geohazards researchers often diverge regarding their approaches to and interests in addressing operational ethics for geohazards research.

Keywords Disaster diplomacy · Disaster risk management · Disaster risk reduction · Research ethics · Science diplomacy

1 Introduction

Researching and publishing on geohazards can be hazardous to researchers and to others, leading to operational ethics concerns about balancing the need for geohazards research and the need for everyone's safety. Potential ethical dangers of geohazards research are often known, identified and accepted. Many institutional processes exist

I. Kelman (✉)
Department of Risk and Disaster Reduction, Institute for Global Health, University College London, London, UK
e-mail: ilan_kelman@hotmail.com

UiT The Arctic University of Norway, Tromsø, Norway

for documenting ahead of time and for trying to mitigate these concerns, notably formal risk assessment and research ethics processes. Yet research indicates how following these processes can sometimes lead to ethical concerns; that is, risk assessment and research ethics processes themselves bring ethical difficulties (Bradley, 2007; Silberman & Kahn, 2011). In any case, adhering to institutional requirements does not necessarily fully account for all operational ethics concerns.

Consequently, it is important to continue investigating the numerous, intertwined operational ethics challenges from geohazards research. The topic here is "operational ethics" rather than "ethics" in order to focus on how geohazards researchers act, rather than covering the centuries of multicultural philosophy on defining and debating the meaning of "ethics" in tandem with schools of thought and frameworks for understanding ethics (e.g. LaFollette, 2013). The scoping of this chapter is raising and mapping out some consequential issues related to operational ethics for researching geohazards. This chapter does not set out to resolve issues, mainly since there is not a single solution or approach which would or could apply to all situations and satisfy everyone, as will become evident through the discussion.

The next section examines some operational ethics of the physical and social hazardousness of geohazards research. Then, ways forward are explored within the framing of science diplomacy for geohazards research based on disaster diplomacy. Conclusions indicate that geohazards researchers suggest different approaches for addressing the operational ethics of geohazards research. The key phrases on which this manuscript is based:

Operational ethics: The application of ethical and moral principles in practice, to influence operations and actions, in this case focusing on carrying out science diplomacy.

Science diplomacy: Conducting science peacefully across political jurisdictions, with examples being data collection, joint publications, and conferences.

Disaster diplomacy: Disaster diplomacy researches how and why all forms of disaster-related activities—including mitigation, prevention, planning, response, and recovery—do and do not influence all forms of conflict and cooperation—including diplomacy, war, negative cooperation, and positive peace.

2 Operational Ethics of the Hazardousness of Geohazards Research

Questions about operational ethics (defined in Sect. 1) in geohazards research appear when conducting or publishing science might change or create dangers to the scientists or to others. Dangers can be created physically and socially.

Physical dangers could mean that a geohazard is involved in deaths, injuries, damage and disruption. The very act of researching could morph a typical environmental phenomenon into a hazard. Mount Unzen, Japan, was erupting in 1991 when scientists, journalists and government officials went to observe it. A huge ash and dust cloud, termed a pyroclastic flow or pyroclastic density current, behaved in a way that was surprising at the time and killed 43 people (Nakada & Fujii, 1993). The desire to observe the geohazard directly in order to understand better its behaviour had placed people in physical danger.

Two years later, in Pasto, Colombia, near the active volcano Galeras, volcanologists held a conference which included an optional field trip to the volcano. With some conference attendees warning of possible signs of an eruption and questioning the value of going near the volcano, several scientists still chose to climb the slopes and go into the crater, with statements about the importance of the scientific work, including collecting gas samples. Some tourists and journalists joined the trip, with a few presuming that the scientists would not put themselves in danger, so the situation ought to be safe.

While people were in the crater and on its rim, Galeras had a small eruption that would typically have been inconsequential. In this instance, six scientists and three tourists died—the first recorded deaths from the volcano—and many more were injured (Bruce, 2001; Cardona, 1997). Without the on-site research, scientists and others would not have been in danger from the geohazard.

Social hazards can be similarly produced. Kazakhstan currently has an authoritarian government (Abishev et al., 2024). The oil and gas infrastructure in many places could be damaged by a geohazard, with avalanches of high concern (Zhangabay et al., 2023), leading to leaks and environmental contamination. Local activist groups seek to publicise environmental issues, leading to research on the topic (Dubuisson, 2024). Many researchers from outside Kazakhstan are legally bound by their countries' and institutions' research ethics protocols, demanding anonymity and confidentiality for research participants. This situation might frustrate local activists who are willing to put their lives on the line, exactly as per the lethal January 2022 political protests (Abishev et al., 2024), ironically started by an increase in fossil fuel prices, by supporting publicity outside of Kazakhstan of the details and specifics of their plight.

Researchers are caught between operational research ethics obligations and wanting their research to help the participants as much as possible. The challenges become acute when "citizen science" is applied, using local residents to collect and perhaps analyse systematic data from their surroundings. The act of setting up monitoring equipment or distributing images, videos and other observations could place the citizen scientists in danger. Career scientists could also get in trouble with their governments. India has permitted out-of-country data sharing of seismic and hydrological monitoring for approximately a decade. Before, unauthorised data distribution was illegal.

Another example of operational ethical predicaments within geohazards research is when investigations reveal that infrastructure is in a hazard zone or an increasingly hazardous zone, perhaps from a newly discovered earthquake fault (Quigley et al., 2010) or because a floodplain has changed from river engineering (Tobin, 1995). Suppressing this science is unethical, as would be not informing people living in, working in, or travelling to the zone about the re-evaluated hazardousness. Providing the updated hazardousness might reduce property values, potentially making it impossible for people to sell in order to move to a less dangerous area, with governments not always willing to step in. In the UK, many homes become uninsurable when their flood risk is updated, with implications for sellers and for buyers trying to obtain a mortgage (Christophers, 2019).

Furthermore, publishing information on geohazards could induce behavioural change, placing people in other dangers, both geohazard and otherwise. Settlements near Svínafellsheiði, Iceland, received a no-build order from 2018 to 2020 due to geohazards, which led to anxiety and detrimental social impacts as the population had to live with the known risks and with little way to improve their situation (Matti et al., 2022). Consider if people living there had been given a fair buyout price and resettlement allowance, then moving to the Svartsengi area, which was evacuated during volcanic eruptions from 2021 to 2024 (Troll et al., 2024). Or they might have moved to the 'big city' of Reykjavík and have to deal with crimes which are rarer in rural areas.

After thousands of people were evacuated from the UK Overseas Territory of Montserrat due to volcanic eruptions starting in 1995, with science supporting the decision-making, many families who moved to England expressed unhappiness at the education their children were receiving (Windrass & Nunes, 2003). Ethically, how do geohazard scientists balance volcanic risks, including long-term breathing in of ash that might lead to cancer (Baxter et al., 1999), with social risks, including degraded education, affecting children over the long term?

Countering disinformation from scientists about geohazards presents another ethical concern. At the moment, the scientific consensus, which could change at any point, is that the frequency of tropical cyclones (cyclones, hurricanes and typhoons) is decreasing due to human-caused climate change, while the storms' intensity is increasing (Chand et al., 2022; Knutson et al., 2020). On a few occasions, anthropologists who have never researched changing geohazards under climate change claimed that the number of these storms is increasing due to human-caused climate change. They were hostile or not responsive to being informed of the scientific consensus, with citations provided. Some eventually altered their statements without acknowledging their initial mistake and without stating that tropical cyclone numbers are accepted as declining (e.g. Jacka & Moore, 2023). Others continue spreading incorrect information and aim to silence scientists pointing out the error.

The anthropologists' approach introduces the social hazard of geohazards research being undermined, because the anthropologists' mistakes and disinformation could be exploited by the concerted campaign to undermine climate change science and to avert action on human-caused climate change (Oreskes & Conway, 2010). When the anthropologists decline to speak to geohazards or climate change scientists, or to properly cite the published research on climate change and tropical cyclones, an operational ethics dilemma emerges between avoiding censorship and harming colleagues by openly challenging scientific disinformation from scientists who specialise in different fields.

3   Science Diplomacy for Geohazards Research

Identifying challenges regarding the operational ethics of geohazards research is generally straightforward. Finding robust approaches for resolving difficulties is much harder. Science diplomacy for the context of disaster diplomacy research offers an avenue of exploration, with science diplomacy and disaster diplomacy defined in Sect. 1. For disaster diplomacy, science diplomacy covers disaster research, including geohazards research.

As noted in the previous section, institutional risk assessment and research ethics processes are one set of formal codes of practice. Others come from scientific societies seeking ethical collaboration and aiming to help bring together researchers and research participants according to a common ethical baseline.

The International Association of Volcanology and Chemistry of the Earth's Interior (IAVCEI) used to have "Safety Recommendations for Volcanologists and the Public" (IAVCEI, 1994), but now has a "Code of Conduct for IAVCEI meetings and events".[1] The latter states as unacceptable behaviour, "Engaging in 'parachute science' or geo-colonialism towards local scientists or the general public during events including field trips". Some institutions retain their own code of conduct for volcanology (e.g. University of Uppsala, 2024). The IAVCEI Subcommittee for Crisis Protocols (1999) proposed "Professional conduct of scientists during volcanic crises", which led to objections (Geist & Garcia, 2000), followed by a response (IAVCEI Subcommittee for Crisis Protocols, 2000) trying to allay the objections.

The long-standing debates over the need for and forms of such documents, alongside the lack of longevity and implementation of specific ones proffered, demonstrate some resistance to codes of conduct for supporting ethical behaviour in geohazards science. Divergence occurs between scientists who see them as useful and those who disagree. Reasons are due to varying levels of risk aversion, dissimilar approaches to research collaboration and distinct prioritisations of risk reduction compared to risk management. These efforts for operational ethics for geohazards research have thus far not yielded science diplomacy, in terms of coming together for agreed, safe and principled research.

Pushing hard for such efforts could backfire, scuttling disaster diplomacy within geohazards research. If scientists see their baseline duty as being the generation and dissemination of new knowledge without simultaneously considering the consequences, then they could consider any ethics suggestions as impositions, which undermine research (cf. Geist & Garcia, 2000; Sluka, 2020). This tension pervades not just geohazards research but also many other fields, including the development of nuclear bombs (Weeramantry, 1985). Roles, duties, desires and rights of scientists to be involved in societal issues beyond the strict remit of contemporary science, steeped in narrow conceptualisations of knowledge production, lead to operational ethical issues on which little convergence is evident for now.

---

[1] Code of Conduct for IAVCEI meetings and events: https://www.iavceivolcano.org/code-of-conduct-for-iavcei-meetings-and-events (accessed 7 February 2025).

Another example comes from the Society for Applied Anthropology (SfAA) in relation to its ethics, resolutions and bylaws. When one SfAA leader publicly attacked this author for supporting ethical and critiquing examination of the implications of human-caused climate change (a geohazard), other leaders declined to investigate detailed documentation of the situation. When follow-ups were not replied to, this author publicised that SfAA's rules lacked statements on bullying and silencing. In response, SfAA leaders removed this author from one email list against their own rules and from two other online venues.

In circumstances in which concerns about bullying and silencing are raised and leaders respond with bullying and silencing, how effective are any operational ethics mechanisms? Consider the Latin adage said to be by Juvenal, "Quis custodiet ipsos custodes?" meaning "Who guards the guards?" It could become "Who leads the leaders?" when they do not conform to the rules they created. The interpretation could be that reconciliation is not of interest, particularly for uniting over verifiable science, scientific processes and research application. Furthermore, monitoring, enforcement and accountability mechanisms are seemingly discouraged for anthropological exchange about geohazards.

From these volcanology and anthropology examples, it would appear that, irrespective of how much we might wish otherwise, geohazards research has politics fundamentally associated with it, simply because people interact and interpret. While ethical concerns were raised about actions, namely seeking operational ethics through codes of conduct and opposing bullying/silencing, not acting raises ethical concerns too. Doing nothing about scientists' conduct is a decision and has ethical implications, as much as trying to do something. In these examples, it is not even clear whether the starting point should be ethics risk reduction or ethics risk management, including with regard to reputational risk, and how to balance riskophobia with riskophilia (cf. Reiner et al., 2001).

However much solid, operationalisable principles are desired and needed for geohazards research across disciplines, practicalities intercede. For instance, the question is not so much how often seismologists check the seismic safety of accommodation they book for conferences and field work. The question is how possible it would be to ever obtain a credible and accurate assessment of the accommodation's seismic safety. What should the scientist do if the building is independently verified as conforming to top earthquake safety design and operation standards, but it sits in a floodplain or lacks adequate fire exits? How important is safe driving compared to seismic safety—and how could both be implemented simultaneously? These are daily life questions for everyone, not just for geohazards research.

When an asteroid is calculated to have a precise probability of hitting the Earth on a specific date (cf. ESA website[2]), should all geohazards researchers outside the possible impact zone plan to avoid travel to places along that path at that time, until more certainty is established about the asteroid's behaviour? If impact is confirmed and the world does not prevent it (Kofler et al., 2019) through science diplomacy for

---

[2] Asteroid 2024 YR4: Latest Updates. https://blogs.esa.int/rocketscience/2025/02/04/asteroid-2024-yr4-latest-updates (accessed 25 February 2025).

disaster diplomacy, then operational ethical dimensions emerge regarding how best to support affected geohazards researchers and everyone else.

Just as principles of disaster diplomacy are not typically observed in practice, ethical principles for geohazards research are not necessarily operationalised in practice. Plenty of experience and efforts exist, but much more could be completed.

4 Conclusions

Geohazards research presents operational ethics challenges, without fully accepted and demonstrated ways forward for managing these situations. Discussions and efforts continue, along with continuous evaluation of all processes. Agreement is not easy and might not be achievable.

Theoretical principles related to operational ethics continue to be espoused, from historical (Fitch & Stuhl, 2024) through to contemporary and future (Peppoloni & Di Capua, 2024) geohazards analyses. In practice, geohazards researchers vary in addressing the operational ethics of geohazards research, indicating the importance of continuing discussion and exchange. Gaillard and Peek (2019) advocate for a code of conduct for research within disaster zones, with Gaillard having led the development and adoption of disaster studies manifesto and accord.[3] Kendra and Wachtendorf (2020) disagree with this approach, and neither is listed as having signed the manifesto or the accord. Thus far, no published research evaluates the manifesto or the accord applied in practice, nor analyses possible operational drawbacks (cf. Bradley, 2007; Silberman & Kahn, 2011), leading to questions about the documents' usefulness and usability.

Aung et al. (2019) suggest that ethics approvals for disaster research should be facilitated and expedited. Sluka (2020) indicates that institutional processes are sometimes designed to undermine needed research in this field. With geohazards research ever-evolving alongside philosophical and ethical logics and articulations, the importance of open and respectful discussion remains paramount, in tandem with not putting people at uninformed or involuntary risk. The 1991 and 1993 deaths of volcanologists and others with them, among many other incidents, led to significant introspection and debates among geoscientists. Operational ethics for researching geohazards means not awaiting fatalities or other detrimental impacts, followed by documenting the catalogue of failures leading to those problems. It means examining and acting on the risks involved in geohazards research before it is too late.

---

[3] Power, prestige and forgotten values: A disaster studies manifesto/Priorities, values, and relationships: A disaster studies accord: https://www.radixonline.org/manifesto-accord (accessed 5 January 2025).

References

Abishev, G., Kurmanov, B., & Sabitov, Z. (2024). Authoritarian succession, rules, and conflicts: Tokayev's gambit and Kazakhstan's bloody January of 2022 (*Qandy Qantar*). *Post-Soviet Affairs, 40*(6), 429–451. https://doi.org/10.1080/1060586X.2024.2377929

Aung, M. N., Murray, V., & Kayano, R. (2019). Research methods and ethics in health emergency and disaster risk management: The result of the Kobe expert meeting. *International Journal of Environmental Research and Public Health, 16*(5). https://doi.org/10.3390/ijerph16050770

Baxter, P. J., Bonadonna, C., Dupree, R., Hards, V. L., Kohn, S. C., et al. (1999). Cristobalite in Volcanic Ash of the Soufriere Hills Volcano, Montserrat, British West Indies. *Science, 283*, 1142–1145. https://doi.org/10.1126/science.283.5405.1142

Bradley, M. (2007). Silenced for their own protection: How the IRB marginalizes those it feigns to protect. *ACME, 6*(3), 339–349. https://doi.org/10.14288/acme.v6i3.782

Bruce, V. (2001). *No apparent danger: The true story of volcanic disaster at Galeras and Nevado Del Ruiz*. HarperCollins.

Cardona, O. D. (1997). Management of the volcanic crises of Galeras volcano: Social, economic and institutional aspects. *Journal of Volcanology and Geothermal Research, 77*(1–4), 313–324. https://doi.org/10.1016/S0377-0273(96)00102-3

Chand, S. S., Walsh, K. J. E., Camargo, S. J., Kossin, J. P., Tory, K. J., et al. (2022). Declining tropical cyclone frequency under global warming. *Nature Climate Change, 12*, 655–661. https://doi.org/10.1038/s41558-022-01388-4

Christophers, B. (2019). The allusive market: Insurance of flood risk in neoliberal Britain. *Economy and Society, 48*(1), 1–29. https://doi.org/10.1080/03085147.2018.1547494

Dubuisson, E.-M. (2024). Whose world? Discourses of protection for land, environment, and natural resources in Kazakhstan. In A. Obydenkova (Ed.), *Sustainable development, regional governance, and international organizations: Implications for post-communism* (pp. 410–422). Problems of Post-Communism, Routledge. https://doi.org/10.1080/10758216.2020.1788398

Fitch, B., & Stuhl, A. (2024). Agnes, revisited: Methods and principles for community-engaged research on historic flood disasters. *Journal of Environmental Studies and Sciences, 14*, 694–709. https://doi.org/10.1007/s13412-024-00976-4

Gaillard, J. C., & Peek, L. (2019). Disaster-zone research needs a code of conduct. *Nature, 575*, 440–442. https://doi.org/10.1038/d41586-019-03534-z

Geist, D., & Garcia, M. (2000). Role of science and independent research during volcanic eruptions. *Bulletin of Volcanology, 62*, 59–61. https://doi.org/10.1007/s004450050291

IAVCEI. (1994). Safety recommendations for volcanologists and the public. IAVCEI. Retrieved June 10, 2025, from https://volcanology.geol.ucsb.edu/safety.htm

IAVCEI Subcommittee for Crisis Protocols. (1999). Professional conduct of scientists during volcanic crises. *Bulletin of Volcanology, 60*, 323–334. https://doi.org/10.1007/PL00008908

IAVCEI Subcommittee for Crisis Protocols. (2000). Reply. *Bulletin of Volcanology, 62*, 62–64. https://doi.org/10.1007/s004450050292

Jacka, J. K., & Moore, A. (2023). Flood and fire: Reorganizing lives around extreme conditions. *Environment and Society, 14*(1), 1–3. https://doi.org/10.3167/ares.2023.140101

Kendra, J., & Wachtendorf, T. (2020). Disaster-zone research: No need for a customized code of conduct. *Nature, 578*, 363. https://doi.org/10.1038/d41586-020-00459-w

Knutson, T., Camargo, S. J., Chan, J. C. L., Emanuel, K., Ho, C.-H., et al. (2020). Tropical cyclones and climate change assessment: Part II: Projected response to anthropogenic warming. *Bulletin of the American Meteorological Society, 101*(3), E303–E322. https://doi.org/10.1175/BAMS-D-18-0194.1

Kofler, R., Drolshagen, G., Drube, L., Haddaji, A., Johnson, L., et al. (2019). International coordination on planetary defence: The work of the IAWN and the SMPAG. *Acta Astronautica, 156*, 409–415. https://doi.org/10.1016/j.actaastro.2018.07.023

LaFollette, H. (Ed.). (2013). *The international encyclopedia of ethics* (9 Volume Set). Wiley.

Matti, S., Ögmundardóttir, H., Aðalgeirsdóttir, G., & Reichardt, U. (2022). Psychosocial response to a no-build zone: Managing landslide risk in Iceland. *Land Use Policy, 117*. https://doi.org/10.1016/j.landusepol.2022.106078

Nakada, S., & Fujii, T. (1993). Preliminary report on the activity at Unzen Volcano (Japan), November 1990-November 1991: Dacite lava domes and pyroclastic flows. *Journal of Volcanology and Geothermal Research, 54*(3–4), 319–333. https://doi.org/10.1016/0377-027 3(93)90070-8

Oreskes, N., & Conway, E. (2010). *Merchants of doubt*. Bloomsbury.

Peppoloni, S., & Di Capua, G. (2024). *Geoethics for the future: Facing global challenges*. Elsevier. https://doi.org/10.1016/C2022-0-00487-6

Quigley, M., Villamor, P., Furlong, K., Beavan, J., Van Dissen, R., et al. (2010). Previously unknown fault shakes New Zealand's South Island. *Eos, 91*(49), 469–470. https://doi.org/10.1029/2010EO 490001

Reiner, R., Livingstone, S., & Allen, J. (2001). Casino culture: Media and crime in a winner–loser societ. In K. Stenson & R. Sullivan (Eds.), *Crime, risk and justice: The politics of crime control in liberal democracies* (pp. 174–194). Willan Publishing. Retrieved June 10, 2025, from https://www.researchgate.net/publication/30529035_Casino_culture_media_and_crime_ in_a_winner-loser_society

Silberman, G., & Kahn, K. L. (2011). Burdens on research imposed by institutional review boards: The state of the evidence and its implications for regulatory reform. *The Milbank Quarterly, 89*(4), 599–627. https://doi.org/10.1111/j.1468-0009.2011.00644.x

Sluka, J. A. (2020). Too dangerous for fieldwork? The challenge of institutional risk-management in primary research on conflict, violence and 'Terrorism.' *Contemporary Social Science, 15*(2), 241–257. https://doi.org/10.1080/21582041.2018.1498534

Tobin, G. A. (1995). The Levee love affair: A stormy relationship. *Water Resources Bulletin, 31*, 359–367. https://doi.org/10.1111/j.1752-1688.1995.tb04025.x

Troll, V. R., Deegan, F. M., Thordarson, T., Tryggvason, A., Krmíček, L., et al. (2024). The Fagradalsfjall and Sundhnúkur Fires of 2021–2024: A single magma reservoir under the Reykjanes Peninsula, Iceland? *Terra Nova, 36*(6), 447–456. https://doi.org/10.1111/ter.12733

University of Uppsala. (2024). *Safety regulations for field work at non glacier volcanoes*. Department of Earth Sciences, University of Uppsala, Uppsala. Retrieved June 10, 2025, from https://www.uu.se/download/18.44c2e1e418f37978f2198ca/1714727890661/ Safety%20at%20volcanoes.pdf

Weeramantry, C. G. (1985). Nuclear weaponry and scientific responsibility. *Journal of the Indian Law Institute, 27*(3), 351–386. Retrieved June 10, 2025, from https://www.jstor.org/stable/439 52248

Windrass, G., & Nunes, T. (2003). Montserratian mothers' and English teachers' perceptions of teaching and learning. *Cognitive Development, 18*(4), 555–577. https://doi.org/10.1016/j.cog dev.2003.09.007

Zhangabay, N., Ibraimova, U., Suleimenov, U., Moldagaliyev, A., Buganova, S., et al. (2023). Factors affecting extended avalanche destructions on long-distance gas pipe lines: Review. *Case Studies in Construction Materials, 19*. https://doi.org/10.1016/j.cscm.2023.e02376

Geoheritage, Geotourism, and Geoethics as Support for Rural Development: A Case Study from Sundarbans Wetlands—A Ramsar Site of West Bengal, India

Sayani Khan⦿, Amit Mandal, and Tarak Nath Khan⦿

Abstract The Sundarbans, the world's largest contiguous mangrove tigerland, is iconic for its biodiversity, geodiversity, and cultural diversity. However, its geoheritage and geotourism have not been properly explored. This study aims to highlight the geotourism potential and geoethical dimensions in promoting rural development in the Sundarbans, particularly in relation to geoheritage and geoconservation. A geosite inventory was created for the first time, and each site was evaluated using the Geosite Assessment Model (GAM) to identify the most suitable locations for future geotourism development and to support rural growth. Effective management and precautionary conservation strategies that integrate both ecotourism and geotourism are essential. Incorporating geoethical principles in geosite management can significantly contribute to sustainable geoconservation.

Keywords Sundarbans Wetlands · Geotourism · Geoheritage · Geoethics · Geosite · Rural development

1 Introduction

World's largest delta system, Sundarbans, is formed by the Ganges–Brahmaputra River confluence in the Bay of Bengal. This tide-dominated estuary is intersected by a mesh-like network of rivers, channels, and tidal creeks (Fig. 1), which supports the world's largest contiguous mangrove belt and the only mangrove tigerland on

S. Khan (✉) · A. Mandal
Geological Survey of India, Eastern Region, Kolkata, India
e-mail: sayanikhan@gmail.com

Present Address:
A. Mandal
Geological Survey of India, Central Headquarters, Kolkata, India

T. N. Khan
Department of Zoology, Maulana Azad College, Kolkata, India

S. Peppoloni and G. Di Capua, *Geoethics and Geosciences Serving Society*, SpringerBriefs in Geoethics, https://doi.org/10.1007/978-3-032-03754-1_6

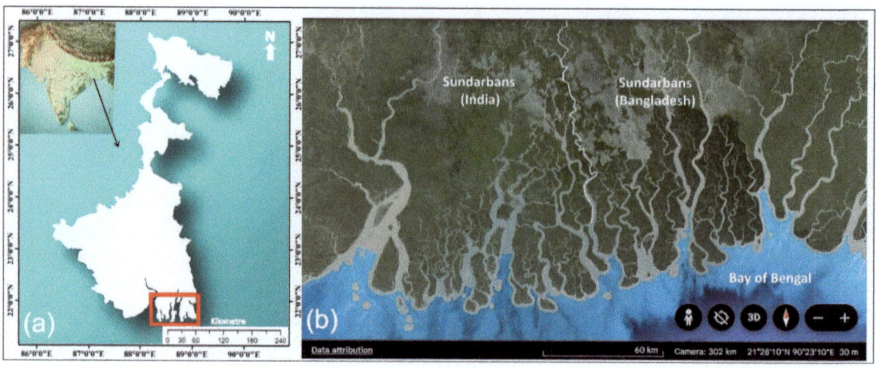

Fig. 1 **a** Sundarbans Wetland in South 24 Parganas District, West Bengal, India. The study area is marked in red box. **b** Sundarbans formed at the confluence of the Ganges–Brahmaputra Rivers with innumerable small channels and creeks in the Bay of Bengal. Google Earth Pro has been used as the base map

earth. Sundarbans has already badged the tags of (1) Project Tiger Reserve (1973), (2) National Park (1984), (3) UNESCO World Heritage Site (1987), (4) Biosphere Reserve (1989), and (5) Ramsar Site (2019).

An integrated approach involving geoscientific observation and ethnographic aspects was made in this endeavor to find potential sectors in Sundarbans for (i) promoting geotourism to derive socio-economic benefits. This is very important for the local communities who are practically living on the edge; (ii) its environmental and cultural preservation, allowing it to grow as a flourishing biosystem; and (iii) education, awareness, and appreciation (Allan, 2015; Brilha, 2016; Brocx & Semeniuk, 2019; Dowling & Newsome, 2018; Gray & Gordon, 2020; Peppoloni & Di Capua, 2023) so that proper care is taken by all concerns to maintain its sustainability. The study also attempts to emphasize the geoethical values and suggest strategies for sustainable geoconservation of Sundarbans.

2 Geological Setting of Sundarbans

Sundarbans consists of quaternary sediments composed of dominantly mud-clay, silt, and fine sand of Bhagirathi and Katwa formations, the environment of deposition being fluvial surrounding the river channels and tidal around the estuarine part. Depositional landforms (within creeks) include tidal flat, mudflat, intertidal marsh lands, creek margins, mid-channel bars, trunk channels/island bars, and point bars. Erosional landforms (within creeks) preserved are runnels, V-shaped channels formed during low tide, erosional embankment landforms like collapsing and arcuate sink of banks, bank erosion, and embankment migration. Depositional landforms (along sea front) include beach—narrow beach face with dunes preserving different wind-borne structures. Erosional landforms (along sea front) like dunes, ripples, beach exposing underlying mud or muddy beach showing mud balls, swash mark, V-swash marks, rhomboid ripple marks, rill marks, low amplitude backwash ripples, laminations, remnants of tree trunks-roots, and submerged remnants of old structures exposed during low tide are also present. In-channel tidal flats, inter-distributary supratidal flats, mudflats and runnels, islands and island channels, mid-channel shoals and bars within major creeks are the common geomorphological set up in the Sundarbans (Ganguly et al., 2006).

3 Diversity of Sundarbans

Sundarbans is a unique biome playing a significant role in the life support system of local and regional people. It preserves rich biodiversity, geodiversity, and cultural diversity (Fig. 2).

3.1 Biodiversity

The Sundarbans harbors more than 300 plant species. Notable mangroves include Sundari (*Heritiera fomes*), Goran (*Ceriops decandra*), Geoya (*Excoecaria agallocha*), Bain (*Avicennia* spp.), Garjan (*Rhizophora apiculata*), and Keora (*Sonneratia* spp.). Many plants have immense ethnobiological values. For instance, Hargoja (*Acanthus ilicifolius*) is used as an effective medicine to treat tiger bites, while Hental (*Phoenix paludosa*) sticks help preventing tiger attacks. Mangroves serve as a barrier against cyclones, tidal surges, and saltwater intrusion (Danielsen et al., 2005) and provide livelihood of the local people (Ewel et al., 1998).

Notable animals include Royal Bengal Tiger (*Panthera tigris tigris*), Fishing Cat ((*Prionailurus viverrinus*), Gangetic (*Platanista gangetica*) and Irrawaddy (*Orcaella brevirostris*) Dolphins, River Terrapin (*Batagur baska*), Greater and Lesser Adjutants (*Leptoptilos dubius and L. javanicus*), Estuarine Crocodile (*Crocodylus porosus*), among others. Around 315 species of birds, ~120 species of fish, a wide variety of reptiles, and innumerable invertebrates enrich its territory.

3.2 Geodiversity

Sundarbans, a classic example of a tide-dominated delta exhibiting active geomorphic processes, is a paradise for geoscience research. Wave-dominated sedimentary regimes in Matla–Saptamukhi riverfronts, asymmetrical ripples near Bakkhali-Henry islands, aeolian dunes, cross-beddings at Sitarampur beach, accretionary tidal landforms in Hatania-Doania channel, mid-channel-bars in the Thakuran River, sandbars at Muri Ganga riverfront, mudflats in the Saptamukhi River are well-preserved.

Fig. 2 Natural diversity of Sundarbans manifesting its ecological–geological–cultural importance and necessity of ecotourism–geotourism. Mangrove forests of: **a** Garjan; **b** Goran and Sundari; **c** Keora; Endangered fauna: **d** Estuarine Crocodile; **e** Greater Adjutant; **f** Bengal Tiger; Geological structures: **g** Mudflat and tidal creek at northwestern edge of Pathatpratima; **h** Erosion and slumping along northern bank of Hatania-Doania Khal; **i** Accretion along eastern bank of Hatania-Doania Khal near Iswaripur village; Cultural celebrations: **j** Traditional Dance in "Bonbibi" Festival; **k** Dance-drama during "Manasa" Festival; **l** Traditional dance during local Folk Festival

River bank erosion is pronounced due to high-energy current flow and wave activity, banks of Sagar–Gorumara–Ghoramara islands in the Hugli–Muri Ganga Rivers being highly susceptible (Das et al., 2013). Mangrove ecosystem morphodynamics and diverse geomorphic features of creeks–channels–rivers enriched with magnificent biodiversity, thus has vast possibilities for developing geoheritage and geotourism site. Preservation of hydrological–atmospheric–oceanographic parameters is significant to fight climate change. River bed morphology and coastal morphology play important roles in maintaining navigability. Moreover, the geological evolution of the Sundarbans, as well as Bengal basin is of utmost importance in global context because of its connection to the Supercontinent (Pangea) Cycle.

3.3 Cultural Diversity

Sundarbans is famous for its cultural heritage. It preserves a combination of human–wildlife conflict and folk culture-religion (Chakraborty, 2005). The "Goddess of Forest"—"Bonbibi" is worshiped here as the guardian of forest and savior of lives from tigers against curse of the "God of Tiger"—"Dakshinrai", in disguise of a tiger. Tigers are revered as divine entities. They gave rise to a population of "tiger widows"—a category of ostracized women. The people penetrate forest to earn a livelihood, inviting frequent tiger-attacked deaths. Their widows are ritually prevented from traditional occupations to avoid such deaths; the wives lead widow-life until the return of their husbands. These women, along with "crocodile widows", suffer severe poverty and hard cursed lives. Snakes take a large toll of lives every year and become a part of the rituals and culture of the region. The Snake-Goddess, "Manasa", is worshiped with songs, dances, and dramas for prevention and cure of snakebite. Sundarbans's Netidhopani has been noted as a spot of snakebite-cure in her myth containing text, *Manasa-mangal*s (sixteenth–seventeenth century). Besides, Sundarbans has its own heritage of folk songs, tribal dance, pot painting, etc.

4 Natural and Anthropogenic Threats to Sundarbans

Sundarbans is under severe threats due to bank erosion, loss of riverine islands and mangroves, unsustainable development, and change in land use pattern, habitat alteration, and loss of biodiversity. Pollution, especially due to untreated sewage from Kolkata Metropolis, including heavy metals from Bantala Leather Complex, natural geomorphic processes, change in hydraulic parameters with other oceanographic and atmospheric attributes, and aid in the threats to Sundarbans. Reduction of fresh water inputs is another threat to Sundarbans, which has already led to a decline in Sundari and Golpata (*Nypa fruticans*) (Bhatt & Kathiresan, 2011).

Climate change is one of the most important environmental issues affecting the Sundarbans through results in increasing temperatures, rising sea level (Hazra et al., 2002) and intensifying the frequency of tropical cyclones. Cyclones like Aila in 2009, Amphan in 2020, and Sitrang in 2022, have wreaked havoc impacts in the region. Sea level rise has already resulted in the submergence of two islands, Suparibhanga and Lohacharra, while many others are facing similar threats.[1] Climate change threatens 33% of the global mangroves (IUCN, 2024). India lost 40% of its mangrove area during the last century (Govt. of India, 1987). Today, more than half of the world's mangrove ecosystems are at risk of collapse (IUCN, 2024). Therefore, it is high time to take better management and precautionary conservation measures on a serious note on geoethical point of view.

---

[1] http://www.thedailystar.net/2006/12/22/d61222011611.htm (accessed 10 June 2025).

5   Objectives of This Study

Present study attempts to (1) prepare an inventory of geosites in Sundarbans for the first time; (2) discover the most suitable geosites for future geotourism development to support rural development; and (3) reveal information about the major areas of improvement for each evaluated geosite. Development of geotourism in Sundarbans will also beneficiate protection and conservation by increasing awareness about nature as well as encourage educational–aesthetic–spiritual–religious–cultural values, knowledge systems, social relations. Geoethical implications in geosite management can contribute to sustainable geoconservation in Sundarbans Wetlands.

6   Methodology

Geosite Assessment Model (GAM) (Vujičić et al., 2011) was applied to assess 20 selected geosites (Figs. 3 and 4). For each geosite, data were collected from 153 people, including residents, tourists, and working staff of Sundarbans through a questionnaire. These data were analyzed using descriptive statistics and the Analytic Hierarchy Process (AHP).

Twenty-seven factors, categorized as Main Value (MV) and Additional Value (AV), were used as assessment criteria. Main Values were subdivided into Scientific/Educational Value (VSE), Aesthetic/Beauty Value (VSA), and Protection Value (VPr). Additional Values were subdivided into Functional Value (VFn), and Tourist/Travel Value (VTr). Each of these values was composed of various criteria (Table 1). A total of 12 MV sub-indicators and 15 AV sub-indicators (Table 1) have been valued/numbered following 5-point Likert scale (Likert, 1932), i.e., from 0.00 to 1.00 (0.00, 0.25, 0.50, 0.75, 1.00, considering 0 as not important at all and 1 as very important), which defines the GAM equation:

$$\text{GAM} = \text{MV} + \text{AV} \ [\text{Where MV} = \text{VSE} + \text{VSA} + \text{VPr}; \text{and AV} = \text{VFn} + \text{VTr}]$$

$$\text{M-GAM} = \text{Im}(\text{GAM}) = \text{Im}(\text{MV} + \text{AV}) \left[\text{Where, Im} = \text{Important Factor}\right]$$

M-GAM stands for Modified Geosite Assessment Model. It adjusts the basic GAM score by applying an Importance Factor (Im). So, the final score depends not just on the total value (MV + AV), but also on how important each component is deemed to be in the given context.

Important Factor was calculated using AHP to find out the comparative priority assessment criteria and acceptability of the model. AHP is a method where ideas are shared among the identified groups to make solutions among themselves, and these ideas are digitized according to Saaty's scale (Saaty, 1980) (Table 2), the problem being shaped in a hierarchical tree structure. A pair-wise comparison between each

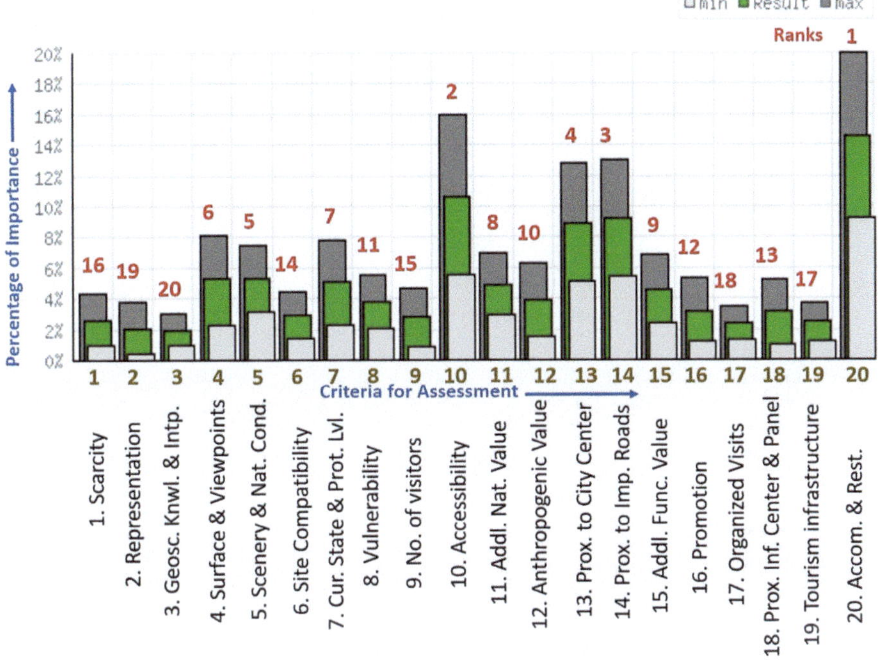

Fig. 3 Consolidated result of AHP Method indicating rank of priority while considering assessment criteria for geosite assessment study

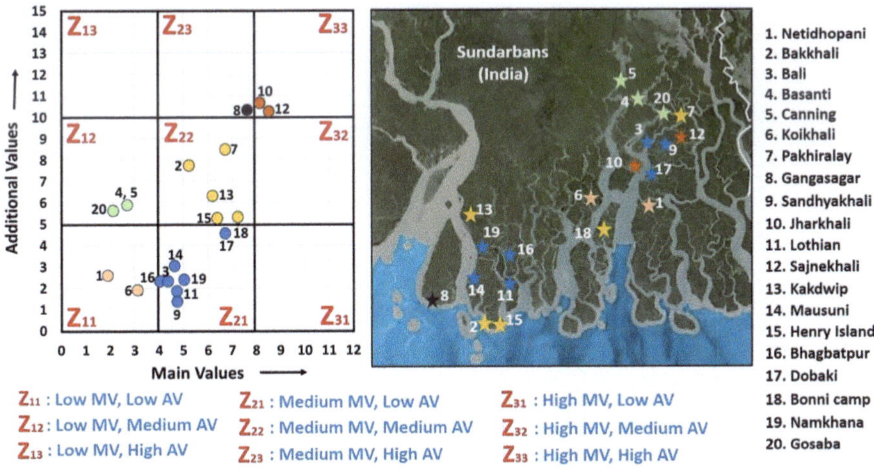

Fig. 4 M-GAM Matrix of geosite assessment inferring Sajnekhali and Jharkhali as the most promising geotourism spots of Sundarbans Wetlands in the present scenario

Table 1 Structure of M-GAM (modified after Dezilia & Harnani, 2023). List of 27 geosite assessment criteria involving main values (MV) and additional values (AV) and their subdivisions. SI in the table stands for "sub-indicator"

	MV	Main value
	VSE	**Scientific/educational value**
1	SIMV-1	Scarcity
2	SIMV-2	Representation
3	SIMV-3	Geoscientific knowledge
4	SIMV-4	Interpretation level
	VSA	**Aesthetic/beauty value**
5	SIMV-5	Viewpoints
6	SIMV-6	Surface
7	SIMV-7	Scenery and natural conditions around
8	SIMV-8	Site/object compatibility with surrounding environment
	VPr	**Protection value**
9	SIMV-9	Current state
10	SIMV-10	Protection level
11	SIMV-11	Vulnerability
12	SIMV-12	Number of visitors
	AV	**Additional value**
	VFn	**Functional value**
13	SIAV-13	Accessibility
14	SIAV-14	Additional natural value
15	SIAV-15	Anthropogenic value
16	SIAV-16	Proximity to the city center
17	SIAV-17	Proximity to major/important roads
18	SIAV-18	Has additional functional value
	VTr	**Tourist/travel value**
19	SIAV-19	Promotion
20	SIAV-20	Organized visits
21	SIAV-21	Proximity of information center
22	SIAV-22	Interpretive panel
23	SIAV-23	Number of visitors
24	SIAV-24	Tourism infrastructure
25	SIAV-25	Geotourism guide
26	SIAV-26	Accommodation
27	SIAV-27	Restaurant

Table 2 Saaty's scale (Saaty, 1980) used for preparation of pair-wise comparison matrix for comparing between each assessment criterion

Scale	Numerical Rating	Reciprocal Rating
Extremely preferred	9	1/9
Very strong to extremely preferred	8	1/8
Very strongly preferred	7	1/7
Strong to very strongly preferred	6	1/6
Strongly preferred	5	1/5
Moderately to strongly preferred	4	1/4
Moderately preferred	3	1/3
Equally to moderately preferred	2	1/2
Equally preferred	1	1

criterion was carried out in the first step. Calculation decides the degree of importance between criteria (for each factor pair) based on the 9-scale of Saaty (higher number indicating chosen factors to be more important). Data were normalized on the pair-wise comparison matrix, and Priority/Weight was determined by calculating normalized Geometric Mean. Finally, the Consistency Ratio (CR), an indicator of the degree of consistency, was determined using the following formula:

Consistency Ratio (CR) = Consistency Index (CI)/Random Consistency Index (RI) [Where, CI = $\lambda_{max} - n/n - 1$, λ_{max} is Maximum Eigen Value and n is total number of criteria; RI for 20 * 20 matrix was taken as 1.63 (Saaty, 1980)].

If CR < 0.01, the result is considered consistent and the model acceptable. Lastly, the M-GAM matrix was prepared using the M-GAM values. Geosite assessment was conducted based on the position of each individual geosite within the M-GAM matrix, according to MV–AV valuation using Z-quarter values.

7 Results

The average MV–AV valuation and GAM calculation for twenty geosites are presented in Table 3. The consolidated results (Fig. 3) show that the most important geosite assessment criterion for the Sundarbans is the availability of accommodation and restaurants. Accessibility of the geosite and proximity to major roads are ranked as the second and third most important criteria, respectively. The number of pair-wise comparisons in the assessment model is 190. The Consistency Ratio (CR) was calculated as 0.086, based on the following formula:

$$CI = \lambda_{max} - n/n - 1 = (22.665 - 20)/19 = 0.140; \text{ and}$$

Table 3 Calculation of GAM value following 5-point Likert Scale for 20 geosites of Sundarbans Wetlands

Geosites		MV-VSE	MV-VSA	MV-Pr	MV	AV-VFn	AV-VTr	AV	GAM
1	Netidhopani	0.25	1	0.75	2	2.5	0	2.5	4.5
2	Bakkhali	1	2	2.25	5.25	2.5	5.25	7.75	13
3	Bali	1.25	1.75	1.5	4.5	1.75	0.5	2.25	6.75
4	Basanti	0.5	1.25	1	2.75	2.75	3	5.75	8.5
5	Canning	0.5	1.25	1	2.75	2.75	3	5.75	8.5
6	Koikhali	0.25	1.75	1.25	3.25	1.25	0.5	1.75	5
7	Pakhiralay	1.75	2.5	2.5	6.75	2.25	6.25	8.5	15.25
8	Gangasagar	3.5	2	2.25	7.75	2.5	7.75	10.25	18
9	Sandhyakhali	1.5	1.5	1.75	4.75	1.25	0.25	1.5	6.25
10	Jharkhali	3	2.5	2.75	8.25	3	7.5	10.5	18.75
11	Lothian	1	1.5	2.25	4.75	1.25	0.5	1.75	6.5
12	Sajnekhali	3.5	2.25	2.75	8.5	2.5	7.75	10.25	18.75
13	Kakdwip	2.25	1.75	2.25	6.25	6	0.25	6.25	12.5
14	Mausuni	1.25	1.75	1.75	4.75	1.75	1.25	3	7.75
15	Henry Island	2.25	1.5	2.75	6.5	2	3.25	5.25	11.75
16	Bhagbatpur	1.25	1.25	1.75	4.25	1.5	0.75	2.25	6.5
17	Dobaki	1.5	2.25	3	6.75	1.5	3	4.5	11.25
18	Bonni camp	2	2.25	3	7.25	1.75	3.25	5	12.25
19	Namkhana	1.25	1.75	2	5	1.25	1	2.25	7.25
20	Gosaba	0.25	1.25	0.75	2.25	2.75	2.75	5.5	7.75

$$CR = CI/RI = 0.14/1.63 = 0.086$$

CR < 1 indicates that the result is consistent and the model is acceptable.

Here, the Principal Eigen Value (λ_{max}) is 22.665, and the Random Index (RI) is 1.63. The number of criteria (n) has been considered as 20 instead of the original 27 assessment criteria, as the error percentage in the AHP method exceeds the acceptable range when using more than 20 criteria. Therefore, similar criteria have been merged into a single criterion.

The M-GAM matrix for geosite assessment indicates that, out of the 20 assessed geosites, two geosites—Sajnekhali (No. 12) and Jharkhali (No. 10)—exhibit both high Main Value (MV) and high Additional Value (AV), making them the most suitable for geotourism under current conditions (Fig. 4; Table 3).

The remaining geosites are categorized as follows:

- One geosite, Gangasagar (No. 8), with medium MV and high AV.
- Five geosites—Bakkhali (No. 2), Pakhiralay (No. 7), Kakdwip (No. 13), Henry Island (No. 15), and Bonnie Camp (No. 18)—with medium MV and medium AV.
- Three geosites—Basanti (No. 4), Canning (No. 5), and Gosaba (No. 20)—with low MV and medium AV.
- Nine geosites—Bali (No. 3), Sandhyakhali (No. 9), Lothian (No. 11), Mausani (No. 14), Bhagabatpur (No. 16), Dobaki (No. 17), and Namkhana (No. 19)—with medium MV and low AV.
- Two geosites—Netidhopani (No. 1) and Kaikhali (No. 6)—with low MV and low AV.

8 Discussion

Mangrove ecosystems are exceptional in their ability to provide essential services, including coastal disaster risk reduction, carbon storage and sequestration, and support for fisheries. Their loss is disastrous for nature and people across the globe (IUCN, 2024). Being the largest global and Critically Endangered Mangrove Ecosystem, Sundarbans deserves special attention.

Despite such importance, its richness in geoheritage, and potential for geotourism development, the Sundarbans has been ignored, and the geoheritage of this area has not been appropriately investigated and proposed for tourism purposes. Conservation measures taken so far have been aimed at biodiversity (focusing exclusively on tigers, fishing cat, and estuarine crocodile). The present study provides an idea of the status of values available in the studied geosites and improvement sectors for potential geotourism sites facilitating rural development. The two most promising geosites with geotourism potential, Jharkhali and Sajnekhali, have already been identified as tourist spots, where annual visitors during the last decade exceeded 0.1 million (Department of Sundarban Affairs, Government of West Bengal). The second most potential geotourist spot, Gangasagar, is an established religiously famous tourist place, claiming 1 billion pilgrims arriving on Sagar Island in 2024. Thus, the present geosite assessment supports the current socio-economic status of the Sundarbans.

Geosite conservation in Sundarbans should involve a holistic approach, embracing geodiversity, biodiversity, and cultural diversity conservation measures. "Ecotourism" and "Geotourism" should be in proper balance to promote tourism as well as protect biosphere. Proper conservation will safeguard the Sundarbans biosystem helping the livelihood of its denizens involving rural development. Geoethical implications in geosite management will no doubt contribute to sustainable geoconservation in the Sundarbans Wetlands.

9 Conclusions

This study presents the first inventory of geosites in the Sundarbans and evaluates the potential for geotourism in the region. Geoethical considerations—including sustainable development, resource management, natural heritage protection, public awareness, ethical decision-making, and the promotion of geotourism in geosite development—are expected to contribute to rural development in the Sundarbans Wetlands. The theoretical and practical implications of this study are outlined below:

(1) The assessment of 20 geosites provides insight into their potential for development as geotourism sites. Among these, Sajnekhali, Jharkhali, and Gangasagar are already established ecotourism destinations. The current assessment criteria for these geosites should be leveraged to further develop geotourism. However, geotourism development must be integrated with ecotourism to ensure a proper eco-geo-balanced approach, promoting overall protection and conservation.

(2) Similarly, the current assessment criteria are expected to serve as guidelines for developing geotourism at other potential geosites (e.g., Bakkhali, Pakhiralaya, Kakdwip, Henry Island, Bonnie Camp, and Dobaki).

(3) The development of eco-geotourism at an increasing number of identified geosites is expected to provide additional livelihoods for local communities living on the forest edge. This may discourage activities such as honey collection, crab hunting, and fishing within the forests, thereby reducing the risk of tiger and crocodile attacks. Consequently, this approach will help minimize human–wildlife conflict, benefiting both wildlife and local people.

The development of an area into a geo- or ecotourism site depends on several factors, including accessibility, surrounding environment, socio-economic status of local communities, and the management strategies employed. This study provides baseline assessment data on the potential of 20 geosites, which will help inform strategies for developing the Sundarbans into a rich and diverse eco-geotourism destination. The goal is to achieve sustainable benefits for local residents, wildlife, and the ecosystem as a whole.

Acknowledgements Authors convey their sincere thanks to Dr. Silvia Peppoloni and Dr. Giuseppe Di Capua, National Institute of Geophysics and Volcanology, Italy, for inviting them to contribute to this volume. Authors are grateful to the Geological Survey of India for facilities, and the Ministry of Mines, Government of India, for providing the necessary funds. Sincere thanks are due to the West Bengal Forest Department, Local Administration, residents, students, teachers, visitors, tour operators, etc., of Sundarbans for their active cooperation to carry out this work. The authors gratefully acknowledge the constructive suggestions of two anonymous reviewers on an earlier version of the manuscript.

References

Allan, M. (2015). Geotourism: An opportunity to enhance geoethics and boost geoheritage appreciation. In S. Peppoloni & G. Di Capua (Eds.), *Geoethics: The role and responsibility of geoscientists*. Geological society (Vol. 419, pp. 25–29). Special Publications. https://doi.org/10.1144/SP419.20

Bhatt, J. R., & Kathiresan, K. (2011). Biodiversity of mangrove ecosystems in India. In J. R. Bhatt, D. J. Macinthosh, T. S. Nayar, C. N. Pandey, & B. P. Nilaratna (Eds.), *Towards conservation and management of Mangrove ecosystems in India* (pp. 1–34). IUCN.

Brilha, J. (2016). Inventory and quantitative assessment of geosites and geodiversity sites: A review. *Geoheritage, 8*, 119–134. https://doi.org/10.1007/s12371-014-0139-3

Brocx, M., & Semeniuk, V. (2019). The '8Gs'—A blueprint for geoheritage, geoconservation, geo-education and geotourism. *Australian Journal of Earth Sciences, 66*(6), 803–821. https://doi.org/10.1080/08120099.2019.1576767

Chakraborty, S. C. (2005). The Sundarbans —Terrain, legends, Gods & myths. *Geographical Review of India, 67*, 1–11.

Danielsen, F., Sørensen, M. K., Olwig, M. F., Selvam, V., Parish, F., et al. (2005). The Asian Tsunami: A protective role for coastal vegetation. *Science, 310*(5748), 643. https://doi.org/10.1126/science.1118387

Das, S., Roy Choudhury, M., Das, S., & Khan, S. (2013). Sediment transport & island change detection: A case study from Sagar Island, West Bengal. *International Journal of Earth Science & Technology, 1*(1), 1–22.

Dezilia, D., & Harnani, H. (2023). Geotourism assessment using the M-GAM method (modified geosite assessment model) Sawahlunto Region, West Sumatra. *Journal of Earth and Marine Technology, 4*(1), 29–40. Retrieved June 10, 2025, from https://ejournal.itats.ac.id/jemt/article/viewFile/4881/3441

Dowling, R., & Newsome, D. (2018). Geotourism: Definition, characteristics and international perspectives. In R. Dowling & D. Newsome (Eds.), *Handbook of geotourism* (pp. 1–22). Edward Elgar Publishing Limited. https://doi.org/10.4337/9781785368868.00009

Ewel, K. C., Twilley, R. R., & Ong, J. E. (1998). Different kinds of mangrove forests provide different goods and services. *Global Ecology & Biogeography Letters, 7*(1), 83–94. https://doi.org/10.1111/j.1466-8238.1998.00275.x

Ganguly, D., Mukhopadhyay, A., Pandey, R. K., & Mitra, D. (2006). Geomorphological study of Sundarban deltaic estuary. *Journal of the Indian Society of Remote Sensing, 34*, 431–435. https://doi.org/10.1007/BF02990928

Government of India. (1987). Mangroves in India: Status Report. Ministry of Environment & Forest, Government of India.

Gray, M., & Gordon, J. E. (2020). Geodiversity and the '8Gs': A response to Brocx & Semeniuk (2019). *Australian Journal of Earth Sciences, 67*(3), 437–444. https://doi.org/10.1080/08120099.2020.1722965

Hazra, S., Ghosh, T., DasGupta, R., & Sen, G. (2002). Sea level and associated changes in the Sundarbans. *Science and Culture, 68*, 309–321. Retrieved June 10, 2025, from https://www.researchgate.net/publication/285639426_Sea_level_and_associated_changes_in_Sundarbans

IUCN. (2024). More than half of all mangrove ecosystems at risk of collapse by 2050, first global assessment finds. Press release, 21 May, 2024. Retrieved June 10, 2025, from https://iucn.org/press-release/202405/more-half-all-mangrove-ecosystems-risk-collapse-2050-first-global-assessment

Likert, R. A. (1932). A technique for the measurement of attitudes. *Archives of Psychology, 22*(140), 1–55.

Peppoloni, S., & Di Capua, G. (2023). The significance of geotourism through the lens of geoethics. In M. Allan & R. Dowling (Eds.), *Geotourism in the Middle East*. Geoheritage, Geoparks and Geotourism (pp. 41–52). Springer. https://doi.org/10.1007/978-3-031-24170-3_3

Saaty, T. L. (1980). *The analytic hierarchy process*. McGraw-Hill.

Vujičić, M. D., Vasiljević, Đ. A., Marković, S. B., Hose, T. A., Lukić, T., et al. (2011). Preliminary geosite assessment model (Gam) and its application on Fruška gora mountain, potential geotourism destination of Serbia. *CTA Geographica Slovenica, 51*(2), 361–377. https://doi.org/10.3986/AGS51303

The Anthropocene Event as a Holistic Foundation for Earth Governance

Emlyn H. Koster, Philip L. Gibbard⬤, and Mark A. Maslin⬤

Abstract Introduced in 2000, the Anthropocene concept developed into a high-profile saga. Its definitional switch from a mid-twentieth-century anthropogenic lake sediment to humanity's cumulative impacts on the Earth System starting in the Late Pleistocene triggered confusion and disagreement. Although the Anthropocene Working Group began as a normal Geological Time Scale investigation in 2009—almost a decade after the term was introduced in an Earth System context—the Group became unconventional in 2015 with a fait accompli approach using a microstratigraphic record of actual and experimental nuclear weapon usage amid two dozen accelerating Earth System and socioeconomic trends. On 4 March 2024, the Group's proposal for an Anthropocene epoch/series on those grounds, prematurely announced to the news media, was rejected by the Subcommission on Quaternary Stratigraphy, a decision ratified by the International Commission on Stratigraphy and International Union of Geological Sciences. This outcome was falsely interpreted by some, including leaders of the former Working Group, as a denial of scientific evidence for anthropogenic climate change. Instead, the Earth's imperiled health elevates the Anthropocene controversy into a far-reaching matter in urgent need of clarification across geoscience, related disciplines, and the public at large. These challenges can be mitigated with a geoethical approach to the Anthropocene Event concept, which enlightens humanity about its estrangement from nature and provides a holistic foundation for Earth Governance.

Keywords Anthropocene · Geoethics · Earth–human ecosystem · Earth Governance

E. H. Koster (✉)
Honorary Professor, Evolutionary Studies Institute, University of the Witwatersrand, Johannesburg, South Africa
e-mail: emlyn.koster@wits.ac.za

P. L. Gibbard
Emeritus Professor, Scott Polar Research Institute, University of Cambridge, Cambridge, UK

M. A. Maslin
Professor, Earth System Science, University College, London, UK

© The Author(s), under exclusive license to Springer Nature Switzerland AG 2025
S. Peppoloni and G. Di Capua, *Geoethics and Geosciences Serving Society*,
SpringerBriefs in Geoethics, https://doi.org/10.1007/978-3-032-03754-1_7

1 Context

In the development of the geological profession since the conclusion of James Hutton (1726–1797) that the Earth shows "no vestige of a beginning, no prospect of an end" (Hutton, 2010), this century's focus on the Anthropocene concept has been transformative. But in terms of environments and climates, the popular version of Hutton's "uniformitarian" principle that "the present is the key to the past" has in many ways ceased to be valid. Arguably more prescient was the conclusion of Alexander von Humboldt (1769–1859) that "humanity and nature are deeply intertwined… that nature would persist in the absence of humanity, but humanity cannot exist without nature" (Jackson, 2019).

From their review of global governance institutions, Lopez-Claros et al. (2020) observed: "Most careful observers of our contemporary landscape would have no difficulty in accepting the claim that we have enabled a period of human evolution characterized by," as Brzezinski (1993) put it, "the velocity of our history and the uncertainty of our trajectory." Outlooks linking the Earth and humanity by Crutzen and Stoermer (2000) and by Crutzen (2002) catapulted the pure to applied transformation of geoscience. Revisiting their 2000 paper, Crutzen and Stoermer (2013) recalled that the idea of the Anthropocene was conceived in "a ferment of concern" as they wondered about a world with nature but no people. When later interviewed about the Anthropocene concept in the context of a pioneering exhibition (Fig. 1), Crutzen anticipated that it "develops into a metaphor about the relationship between nature and humankind" (Schwägerl, 2015).

After the Anthropocene Working Group (AWG) formed in 2009 (Zalasiewicz et al., 2017/2018), Koster (2011) opined: "Geology uniquely brings big time and space perspectives to the planning table as an essential frame of reference… The Anthropocene presents geology's best chance to take its rightful place as a core

Fig. 1 Advertising icon for the 2014–2016 exhibition "Welcome to the Anthropocene: The Earth in Our Hands" (Möllers et al., 2015) at the Deutsches Museum in partnership with the Rachel Carson Center for Environment and Society. Courtesy of the Museum's Head of Research

contributor to the harmonious, multidisciplinary pursuit of positive consequences in the world." Mahli (2017) observed: "the Anthropocene has spilled out of its natural sciences origins to become a cultural zeitgeist… about how to understand and respond to human domination of the Earth."

Referring to the Anthropocene, the primary influencers behind the UN's 2015–2030 "Transforming our World" plan[1] opined that "our ultimately short-sighted approach to growth should be enough to give us pause and force a reflection not just of the scale of what we have done but more so of what we need to do now" (Caballero & Lōndono, 2022). In the same vein, Koster (2022) remarked: "Today, it does not seem premature to conclude that the Anthropocene, with all that this term can usefully encompass, looms as the most consequential alteration in the course of history." Boivin et al. (2024) similarly remarked: "The wide cultural currency of the Anthropocene term offers a rich opportunity to engage a global audience with issues that are relevant to us all."

2 Disagreement

On 4 March 2024, a rejection by the Subcommission on Quaternary Stratigraphy (SQS) of the AWG's proposal for an Anthropocene epoch/series with its base marked by peak fallout in 1952 from mid-twentieth-century atomic bomb tests became ratified and announced by the International Commission on Stratigraphy (ICS) and International Union of Geological Sciences (IUGS).[2] International news coverage caused confusion about the status and future of the Anthropocene concept across academia and society. The voting result (Voosen, 2024; Witze, 2024) was assailed by protagonists of the epoch/series approach. For example, Hadly and Barnosky (2024) exclaimed: "it doesn't matter that the high courts of geology recently denied the existence of the Anthropocene Epoch, it's here, and it gets more real every day." Wrongly implying that the Earth's first major anthropogenic changes were in the mid-twentieth century and that the Anthropocene term was no longer of any value, Ghosh (2024) cryptically titled his article "A fond farewell to the Anthropocene" and defiantly remarked: "In the aftermath of the IUGS decision, the policy community now has an opportunity to break from this overreliance on official scientific consensus." A film for television about the Anthropocene (Arte, 2024) was entirely about the epoch idea as if it became the consensus outcome. Reflecting on the AWG's proceedings in Ware (2024), Erle Ellis remarked: "What I saw happening is that it ended up in this very narrow track of breaking Earth's history into two parts. A part that is considered to be transformed, and a part before it where things are natural or untransformed. That narrative is regressive and harmful politically."

---

[1] https://sdgs.un.org/publications/transforming-our-world-2030-agenda-sustainable-development-17981 (accessed 10 June 2025).

[2] https://stratigraphy.org/news/152; https://www.iugs.org/_files/ugd/f1fc07_40d1a7ed58de458c9f8f24de5e739663.pdf (accessed 10 June 2025).

Zalasiewicz et al. (2024) anticipated that supporters will not back away and anticipated that "Even though geologists have rejected the designation of an Anthropocene epoch, the idea of a major planetary transition in the mid-twentieth century remains useful across physical and social sciences, the humanities and policy." In a similar vein, Damianos (2024) commented on "the social construction of geological truth" and "transitioning geology from a descriptive science about the past to a site of warning." Edgeworth (2025) pointed out that a focus on technofossils by Gabbott and Zalasiewicz (2025) wrongly associated them only with the notion of an Anthropocene epoch since the mid-twentieth century. And with an unseemly tone for a scientific debate, Barnosky and Hannibal (2025) portrayed the annoyance of pro-epoch supporters like themselves as "savage academic infighting." The former AWG's strategies also include projecting "the likely future extent and duration of the proposed Anthropocene epoch" (Summerhayes et al., 2024), ignoring the contrary proposal of an Anthropocene Event (McCarthy et al., 2025), and then referring to an Anthropocene epoch with a small "e" (McCarthy et al., 2025). Krause and Trappe (2025) misleadingly stated: "The geological epoch that has seen us transform our planet and its protective layer… is nowadays referred to as the 'Anthropocene… It is about to replace the Holocene." Skelton and Noone (2025) attempted a logic-based argument that the Anthropocene epoch idea is stronger than the Holocene Epoch decision.

Should the Anthropocene epoch/series idea and/or its Crawfordian stage/age idea resurface and again focus on atomic bomb-testing fallout, it will be helpful that principled concerns about the AWG's intention (e.g., Subramanian, 2019; Waters et al., 2015) were already documented. This paper builds on the views of Gibbard et al. (2022), Koster (2022), Ellis (2023, 2024), Koster et al., (2023, 2024), Walker et al. (2023), Edgeworth et al. (2024), Edgeworth (2025) and Bourzac (2025) that the Anthropocene term refers to human impacts on the Earth System, not those just since a specific historic timeline.

Sagan (2013) cautioned: "The suppression of uncomfortable ideas… is not the path to knowledge and it has no place in the endeavor of science." Mindful too of the whole-truth-and-nothing-but-the-truth principle of the legal profession, this paper illuminates understated aspects of the Anthropocene's unconventional genesis and significant broad potential. We do so in the spirit of those who have urged that differing viewpoints be aired.

3   As an Epoch, an Aberrant Approach

The former AWG's epoch approach is deemed aberrant because it neither complied with the conventional approach to new interval development in the Geological Time Scale (GTS) nor with the IUGS Geoethical Promise, which begins: "I will practice geosciences being fully aware of the societal implications… I will put the interest of society foremost in my work."[3] For the Anthropocene, because of its human raison d'être with an ongoing associated chapter of time, these concerns are intertwined. When resigning from the AWG, Ellis (2023) remarked that "to systematically ignore the overwhelming evidence of Earth's long-term anthropogenic transformation isn't just bad science, it is bad for public understanding of the causes of these changes and for action to address them." Ellis (2024) added that the Anthropocene's "recent date and shallow depth are too narrow to encompass the deeper evidence of human-caused planetary change."

Koster et al., (2023, 2024) explained that their rationale to uncouple Anthropocene research from nuclear warfare was not to question the correlational precision of quasi-isochronous fallout from over 2,000 atomic bomb tests. As explained below, their reasoning was that unemotive references to humanity's most horrendous act should be morally unacceptable to geoscience (Koster, 2022). They also clarified that their stance of uncoupling GTS and Archaeological Time Scale events and deposits was because the former arose from natural deep-time stratigraphic evidence (Edgeworth et al., 2019).

Atomic bombing of Hiroshima and Nagasaki killed an estimated 145,000 and 78,000 people, respectively. The gravity of today's ongoing nuclear threat was captured in this reflection: "The atomic bomb created the conditions of contingent catastrophe, forever placing the world on the precipice of existential doom. But in doing so, it created a philosophy of acceptable cruelty, worthy extinction, legitimate extermination. The scenarios for such programmes of existential realization proved endless. Entire departments, schools of thought, and think tanks were dedicated to the absurdly criminal notion that atomic warfare could be tenable for the mere reason that someone (or some people) might survive. Despite the relentless march of civil society against nuclear weapons, such insidious thinking persists with a certain obstinate lunacy" (Kampmark, 2023). When *The New York Times* provided the context to last survivor stories and described a planned World War II exhibition at the Smithsonian Museum with a spotlight on the two bombed cities, it opined that "it's time for the next generation to bear witness and demand change" (Kingsbury et al., 2024). It was therefore alarming that "A Nuclear Anthropocene" without any caveat about pairing these words was among lectures to students enrolled in the 2025 Vienna Anthropocene Network.[4]

As documented by Koster (2022), the AWG described its conclusion to define the base of an Anthropocene epoch/series with a "seemingly unemotive, almost clichéd, use of "bomb spike"" adding that it did not consider itself publicly accountable for research decisions. Those statements were made during an AWG workshop in Berlin in May 2022. At its next workshop in Florence in September 2022, the "bomb spike" was instead referred to with plutonium-centered, multi-proxy language. In a further dilution of the AWG's initial "bomb spike" characterization, Turner and Waters (2024) referred to "a glut of stratigraphic markers... not least, the unique

---

[3] https://www.geoethics.org/geopromise; https://www.geoethics.org/_files/ugd/5195a5_b200d6 ed6d8545c38f8fd15246b2906d.pdf (accessed 10 June 2025).

[4] https://anthropocene.univie.ac.at/ (accessed 10 June 2025).

historical, geopolitical circumstances of the 1950s resulted in above-ground thermonuclear bomb testing introducing radionuclide markers near synchronously into global stratigraphic successions."

Referring to the origination and classification of a GTS interval, Bourzac (2025) summarized that scientists "first form a working group to develop the argument and gather evidence." In the AWG's case, she opined and we agree, as substantiated below—it worked backward with a top-down approach. Whereas approved intervals specify their distinctive content, the Anthropocene epoch idea focused on the particularities of a base, finalized in the AWG's proposal as 1952 AD. More strangely, it arose nine years after the AWG announced its expectation (Waters et al., 2015) and five years after establishing its course through binding votes (Subramanian, 2019), both as a fait accompli centered on the legacy of humanity's most abhorrent invention seven years before its search for evidentiary Global Boundary Stratotype Section and Point (GSSP) sites (Waters et al., 2023). Moreover, the *Bulletin of the Atomic Scientists*, a journal co-founded by Albert Einstein and University of Chicago scientists who helped to develop the first atomic weapons in the secret Manhattan Project[5] and created the Doomsday Clock,[6] was an odd, unfamiliar-to-geoscience choice to promote the Anthropocene epoch idea. In their title, Waters et al. (2015) rhetorically asked, 'Can nuclear weapons fallout mark the beginning of the Anthropocene Epoch?'; they answered that "nuclear sciences are likely to be crucial to the definition of the Anthropocene" and then clarified that "fallout from nuclear weapons appears most suitable" Not only would this illogical order be unconscionable elsewhere in the GTS, the AWG accepted without the customary diligence of stratigraphic research, the emotive original view of Crutzen (2002) that an epoch would be the Anthropocene's appropriate rank. Furthermore, Koster et al. (2023) noted that Waters et al. (2015) anticipated that "a site to define the Anthropocene… would ideally be located between 30 and 60 degrees north of the Equator within undisturbed marine or lake environments." Finalists of the AWG's 12-candidate GSSP site selection process were lakes at 42.2°N and 43.5°N, that is, mid-latitude sites expected to have the most fallout. Then, in a 9-year-delayed refutation of the seven challenges by Finney (2014) to the AWG's approach, Zalasiewicz et al. (2023) countered that the Anthropocene was "conceptualized prior to the search for its chronostratigraphic definition." Indeed, it was, in astonishing detail!

Damianos (2024) described and opined on a further instance of the AWG proceeding in an unconventional manner. Referring to the STRATI conference in Lille, France, in July 2023, which he attended, "Chair Colin Waters announced, pausing for suspense, that the GSSP candidate for the Anthropocene Series/Epoch would be in Crawford Lake, Canada… An announcement had been made as if the GSSP had been decided, when it had yet to be approved by the decision-making committee of the SQS and ICS. The hastiness of the AWG was perceived as an attempt to subvert the procedures associated with amendments to the GTS through mass-media news."

---

[5] https://en.wikipedia.org/wiki/manhattan_project (accessed 10 June 2025).

[6] https://en.wikipedia.org/wiki/doomsday_clock (accessed 10 June 2025).

Before the AWG was created, Zalasiewicz (2008) mused: "Whatever we as a species do from now, we have already left a record that is now indelible, even while the scale of this fossilization event is still in question, and within our power to determine. Humankind has, through its various activities, done enough to preserve its relics into the far future." But the dilemma facing science and society is far worse. As Bird (2023) lamented: "It is my hope that Christopher Nolan's stunning new film on Oppenheimer's complicated legacy will initiate a national conversation not only about our existential relationship to weapons of mass destruction, but also the need in our society for scientists as public intellectuals." The UN Secretary-General's congratulatory statement to the 2024 Nobel Peace Prize-winning activist Japanese organization of Hiroshima and Nagasaki survivors affirmed: "Nuclear weapons remain a clear and present danger to humanity, once again appearing in the daily rhetoric of international relations."[7] Yet epoch-favoring papers by former AWG members have relentlessly continued to ignore any ethical conscience tied to nuclear warfare.

4 As an Event, a Critical Transition

Events and intervals are an important distinction in the Earth's history. Compared to formally defined GTS intervals with their isochronous boundaries, events typically straddle more than one formally defined time interval and are diachronous. Finney and Edwards (2016) observed that the Anthropocene and Renaissance have similar characteristics: "… both refer to richly documented, revolutionary activities." Interestingly, on the heels of the AWG's creation, a global news source featured a corroborating scientific opinion on the Anthropocene: "The edges of historical eras tend to be fuzzy. It would be nice to think that someone awoke in Florence, Italy, one day in the late 1300s—perhaps as spring started—and said, 'Today the Renaissance begins!' We can be sure no one did, if only because historians discern such eras only in retrospect. The same is true of geological epochs. Humans existed when the Pleistocene ended and the Holocene began, 11,500 years ago" (The New York Times, 2011). A historian later affirmed that events are "created partly in hindsight, shaped by processes of retrospection and commemoration" (Lyon, 2017).

Finney and Gibbard (2023) clarified that the Anthropocene Event needs no GSSP, no fixed formal start date, no approval, nor ratification by the ICS and IUGS. Ellis (2024) positioned it as "a complex, transformative, and ongoing event analogous to the Great Oxidation Event" about two billion years ago. Although that geochemical development is the most cited analog to the Anthropocene Event, the evolution of plants, reptiles, and mammals is more publicly relatable to the evolution and behavioral developments of *Homo sapiens*. Zerkle (2024) described the Great Oxidation

---

[7] https://apnews.com/article/nobel-peace-prize-40c351ed629bd515675a8283c811bc38 (accessed 10 June 2025).

Event as "a critical transition in our planet's history," a characterization also befitting of the Anthropocene Event with its Earth System changing features.

By encompassing the cumulative impacts of humans since the Late Pleistocene (e.g., Erwin, 2024), the unfolding Anthropocene Event is a clarion call for multiple Crutzen-styled interventions across what Koster et al. (2024) referred to as the "Earth–Human Ecosystem." Compared to the "Earth System," it valuably conveys the duality of natural and human forces with humanity's obligation to think and act ecologically. By 2020, Dictionary.com had added "existential" to our vocabulary to refer to "grappling with a sense of survival of our planet, loved ones, our ways of life." In addition to an escalation of violent outbreaks and humanitarian disasters, the pandemic-struck world was anxious about anthropogenic warming of climates, melting of glaciers and ice sheets, rising of oceans, and atrophying of ecosystems.

The AWG's highlighting of the post-World War II "Great Acceleration" (Steffen et al., 2015) only relates to the most recent part of the Anthropocene Event, which is a commonly overlooked point. A long view of the Anthropocene illuminates that society seldom seems to grasp, namely that our ancestors have long intruded on natural history in multiple ways. Ten or more human species used to exist: the last one before us, *Homo neanderthalensis* (Bleiberg, 2024), existed until approximately 40,000 years ago. *Homo sapiens* have been moving, mixing, and diversifying since their ancestors migrated out of Africa. As Cajete (2000) noted: "Five centuries ago Europeans arrived on the American continent, but they did not listen to the people who had lived for millennia in spiritual and physical harmony with this land." Koch et al. (2024) concluded: "The Great Dying of the Indigenous Peoples of the Americas resulted in a human-driven global impact on the Earth System in the two centuries prior to the Industrial Revolution."

Davis (2009) summarized today's humanity: "Together the myriad of cultures make up an intellectual and spiritual web of life that envelops the planet and is every bit as important to the well-being of the planet as is the biological web of life that we know as the biosphere. You might think of this social web of life as an 'ethnosphere', a term for the sum total of all the thoughts and intuitions, myths and beliefs, ideas and inspirations brought into being by the human imagination since the dawn of consciousness." He added evocatively that the ethnosphere "is the embodiment of our hopes, the symbol of all we are and that we, as a wildly inquisitive and astonishingly adaptive species, have created."

Another science-and-society benefit of the Anthropocene Event approach is that geology, archaeology, and history can be addressed as a continuity (Fig. 2). Edgeworth, Richter, Waters, and Price (2015) delved into the diachronous and geometrically complex separation of natural geological deposits and humanly modified ground. The "archaeosphere" contains the fossil record of both Indigenous and Western (named for the early cultural dominance of western Europe) People. Couching the Anthropocene as a new worldview and urging reconsideration of a rigid definition to embrace transdisciplinary perspectives, Henderson and Vachula (2024) stressed the psychological value of a "holistic identification of human impacts to the Earth System."

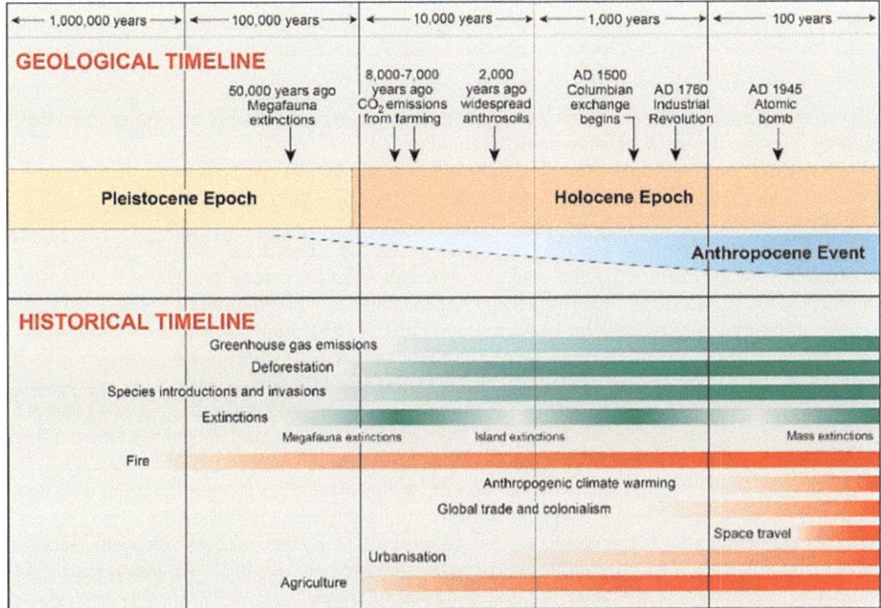

Fig. 2 The Anthropocene Event with its geological and historical contexts slightly modified with permission from Gibbard et al. (2022) by Koster et al. (2023)

While it is also important to understand that humans are not intrinsically different from other organisms in potentially modifying their environment, a major difference is that *Homo sapiens*, with their astonishing cultural diversification of lifestyles, attitudes, and behaviors, became extensively fractured with contested outlooks (Koster et al., 2022). In 2015, the UN General Assembly unanimously embarked on a 15-year "Transforming Our World" plan with 17 Sustainable Development Goals (SDGs) (Caballero & Londõno, 2022; UN, 2015). However, the mid-term report (UN, 2023) stated: "It's time to sound the alarm… At the mid-way point of our way to 2030, the SDGs are in deep trouble… A preliminary assessment of the approximately 140 targets with data show only about 12% are on track; close to half, though showing progress, are moderately or severely off track and some 30% have either seen no movement or regressed below the 2015 baseline… the potential for science, technology and innovation to be applied to the SDGs is vastly untapped and institutional." Although in 2024 the UN announced a new approach to integrate efforts of three SDG clusters, including nature-based solutions,[8] early 2025 reporting revealed "almost no change" in SDG progress.[9]

---

[8] https://integratesdgs.org/ (accessed 10 June 2025).

[9] https://sdg.iisd.org/news/unece-report-reveals-almost-no-change-in-sdg-progress-from-2024-2023/ (accessed 10 June 2025).

Table 1 A sample from almost 300 online comments by readers of "For Planet Earth, this might be the start of a new age" in *The New York Times* (Zhong, 2022): this was also used in Koster et al. (2024)

"Given that it appears that our own era might be rather short, we probably need to get on with naming it before the onset of our own extinction."

"The key question is not did we change things and when, but what will the next few centuries/ millennia look like as a result of our actions and what can we do to minimize adverse consequences."

"Humans with their big brains and big egos have a fatal flaw. They have escaped natural interaction with the Earth. We have insulated ourselves from the natural process of interdependency with nature that historically had balanced any population."

"Nature is dying at our hands. Our beautiful blue Earth is in the intensive care unit, mostly past the point of being saved, due to the stubborn failure of humans to admit that we are the most destructive force to hit the planet since the asteroid destroyed the dinosaurs."

"I like the suggestion to refer to what we're doing as an 'event' and not classify it as a geologic period, whether age or epoch. We are simply too close to the onset, and the changes are coming too fast and too furious to know how long we'll be lasting in this mode… A little-noticed problem with wide adoption and formalization of 'Anthropocene' as an epoch is an unintended flattering of our species' vanity, and a bolstering of our complacency. Telling ourselves that our current mode of choices is a geologic age just solidifies the ethos we need to overcome. Better to recognize our situation as The Anthropic Event. It's a more behavioral classification, which is more accurate, and more amenable to turnaround and recovery."

"We are in this mode as part of an ethos of dominance and control over nature and we are not going to survive this mode without changing that ethos to one of felt connection with and lived reverence toward the more than human world."

"Looking at the grand sweep of this history, the evolution of life on this planet, the organization of sentience in the human mind, the emergence of the complex forms of consciousness, awareness and self-conceptions, it's so painful to look at what we have done with the genius of our species."

Surely all Anthropocene research should be cognizant of the Earth's anthropogenic crises. The high level of alarm evident in reader feedback to an extensive update on anthropogenic impacts in 2022 by *The New York Times* (Table 1) is repeated by a World Economic Forum Global Risks Report with projections for 2027. It rated the worst environmental risks as extreme weather events, biodiversity loss, ecosystem collapse, critical changes to Earth systems, and natural resource shortages.[10] Of note in this regard, Wagreich et al. (2025) misrepresented Koster et al. (2024) by attributing to them a view that "a geoethical stance may do away with the Anthropocene as a useful concept." Absolutely to the contrary, we view a geoethical premise as imperative in all Anthropocene thinking.

---

[10] https://www.weforum.org/press/2025/01/global-risks-report-2025-conflict-environment-and-disinformation-top-threats/ (accessed 10 June 2025).

5   Commentary

Almost a decade ago, a political economist observed: "The Anthropocene is a key theme in contemporary speculations about the meaning of the present and the possibilities for the future… How the Anthropocene is interpreted, and who gets to invoke which framing of the new human age… matters greatly for both the planet and for particular parts of humanity" (Dalby, 2016). Building on this theme, the call to action in the subtitle of Bjornerud's (2018) book on timefulness—"thinking like a geologist can help save the world"—emphasizes that awareness of the Earth's temporal rhythms is critical to our planetary survival. Swyngedouw and Maslin (2024) concurred that the Anthropocene has "opened up fertile ground for interdisciplinary advances on crucial planetary issues." In this vein, a holistic lens on the Anthropocene was valuably recommended by Żuk and Żuk (2024): "We suggest treating [it] not as an abstract idea in the geological and natural sciences, which is difficult for common people to understand, but as a phenomenon that brings political and economic consequences in the real social world."

Classically, the expertise of geoscience focused on interpretation of in-the-ground evidence in and on a pristine Earth. As its focus expanded from the Earth to Earth System Science and as the Anthropocene blurred the past and present into a bigger picture, this profession became well positioned to recognize the vectors and rates of anthropogenic changes in each subsystem as different and quicker than at any previous time in Earth history and in total as an unbreakable continuum.

The preface to "The Challenge of the 21st Century" in a compendium about the emergence of global institutions (Lopez-Claros et al., 2020) was sobering: "We have organizations for the preservation of almost everything in life that we want but no organization for the preservation of mankind. People seem to have decided that our collective will is too weak or flawed to rise to the occasion." Concerned with the ethical precept of longtermism, MacAskill (2022) mused: "To be alive at such a time is both an exceptional opportunity and a profound responsibility: we can be pivotal in steering the future onto a better trajectory. There's no better time for a movement to stand up, not just for our generation or even our children's generation, but for all the generations yet to come."

The responsibility of geoscience to the whole of society requires that its Anthropocene term becomes as well known as, for example, COVID-19 (Fig. 3). Behind the attention to a warming climate and extreme weather is the World Meteorological Organization (WMO). The WMO describes itself as the UN's "authoritative voice on the state and behavior of the Earth's atmosphere, its interaction with the land and oceans, the weather and climate it produces and the resulting distribution of water resources." Earth System-scale research and advocacy knows that this purpose cannot be isolated from integrated research and advocacy about other subsystems. Accordingly, the IUGS should seek to emulate the WMO as a UN consultative science-and-society body.

Johan Rockström, Founder of the Stockholm Resilience Center and Director of the Potsdam Institute for Climate Impact Research in Germany, was among *TIME*

Fig. 3 Symbolism of the core influence that the Earth–Human Ecosystem in an Anthropocene context should be in Earth Governance. Courtesy of VistaCreate

magazine's 100 most influential people in 2023.[11] Since 2009, he has pioneered and communicated the concept of planetary boundaries as environmental guidelines to keep humanity safe. *TIME* also reported that "new business models for shaping the path forward, guiding leaders on how to turn complex science into clear, quantifiable actions" are needed. Breakthroughs in governing the Earth's polycrisis[12] hinge on holistic principles with SMART goals, defined as specific, measurable, assignable, realistic, and time-related,[13] each with an influential champion. Looking back, the UN's 1986 Montreal Protocol to phase out dangerous ozone-depleting gases in the stratosphere stands as exemplary, with Nobel laureate Paul Crutzen its lead researcher.

In the same year as the AWG anticipated a stratigraphic "bomb spike" for the base of its anticipated epoch (Waters et al., 2015), Hamilton and Grinevald (2015) noted that "the Anthropocene represents a radical departure with all evolutionary ideas in human and Earth history." And recently, Bohle et al. (2025) deemed that "the case of the AWG demonstrates the possibilities and limitations of interdisciplinary cooperation among diverse academic disciplines addressing the transformative impact of the Anthropocene concept towards a post-normal framing of the Earth Sciences" (Koster, 2025b). In a similar vein, Bourzac (2025) reprised this ICS statement: "Despite its rejection as a formal unit of the Geologic Time Scale, the Anthropocene will nevertheless continue to be used not only by Earth and environmental scientists, but also by social scientists, politicians and economists, as well as by the public at large. It will remain an invaluable descriptor of human impact on the Earth system."[14]

Writing for the public, Brannen (2018) conjectured: "…if we're to endure as a civilization, or even as a species, for anything more than what might amount to a thin layer of odd rock in some windswept canyon of the far future, some humility is in order about, thus far, our infinitesimal part in the history of the planet." While the

---

[11] https://time.com/collection/100-most-influential-people-2023/6269887/johan-rockstrom/ (accessed 10 June 2025).

[12] https://en.wikipedia.org/wiki/polycrisis (accessed 10 June 2025).

[13] https://en.wikipedia.org/wiki/smart_criteria (accessed 10 June 2025).

[14] Joint statement by the IUGS and ICS on the vote by the ICS Subcommission on Quaternary Stratigraphy, 20 March 2024. https://stratigraphy.org/news/152 (accessed 10 June 2025).

human slice of the Earth's 4.6-billion-year history is indeed tiny, as Edgeworth et al. (2024) pointed out, the Anthropocene is much more than a time interval.

6 Conclusions

Communicating the centrality of geoscience in tackling global environmental and climate exigencies relies on efforts in as many science-and-society networks as possible. This calling is not new. Koster (2020) noted that "Regardless of its eventual formal or informal standing… the term Anthropocene has become valuable shorthand for recognizing humanity as the dominant species which, in a geological nanosecond, has extensively detached itself from the Earth System, endangering the future of both," adding that the entire geological profession should engage with the AWG's work "… to coalesce its activities with those of other disciplines concerned with environmental health and linked human health challenges." Recent examples include Mobilizing an Earth Governance Alliance (Koster, 2024) and Citizens for Global Solutions (Koster, 2025a).

Overall, the geoscience profession should strive to draw attention to the Earth's changes, both those that are natural and those that are human-caused. As an inspiration, a 1962 speech by U.S. President John Kennedy referred to the part that all humans play in the Earth's relentless movement of its primordial water. Addressing crews competing for the America's Cup: "It is an interesting biological fact that all of us have in our veins the exact same percentage of salt in our blood that exists in the ocean, and therefore, we have salt in our blood, in our sweat, in our tears. We are tied to the ocean. And when we go back to the sea—whether it is to sail or to watch it—we are going back from whence we came" (Kennedy, 1962).

Looking forward, it would be most advantageous for both science and society to refer simply to the "Anthropocene" when alluding to humanity's intensifying impacts on the natural subsystems of the Earth System. Accelerated climate changes, which dominate public concerns and policy debates, would be helpfully viewed as an anthropogenic phenomenon in an Earth–Human Ecosystem context. And for already discussed reasons, the rejected unduly restrictive Anthropocene epoch/series concept should slide into obscurity, and the flexible "Anthropocene Event" should describe the current critical transition in Earth history and become the holistic past–present–future framework for crucial efforts toward Earth Governance. In these key respects, a distinction by Christian (2019) between "sharply focused research and unifying intellectual frameworks" is instructive. He pointed out that "good education and research depend on a balance between detail and generality… that help us make sense of, and find meaning in, detailed research." Epoch and event scale thinking about the Anthropocene are, respectively, examples of detailed research and a unifying framework.

Acknowledgements The co-authors commend the vision behind the International Association for Promoting Geoethics (IAPG) and express appreciation to an anonymous reviewer of the pre-final stage of this paper. The first author thanks Simon Turner, Colin Waters, and Jan Zalasiewicz for inviting his participation in the former AWG's workshops in Berlin and Florence, as well as Michael Wagreich for inviting his participation in UNESCO-IGCP Project 732 Language of the Anthropocene meetings in Vienna, Nairobi, and Xi'an.

References

Arte. (2024). *Anthropocene. The Undeniable Truth.* Retrieved June 10, 2025, from https://www.terranoa.com/en/one-off/anthropocene-the-indeniable-truth-2742

Barnosky, A., & Hannibal, M. (2025). *Despite official vote, the evidence of the Anthropocene is clear.* Yale Environment 360, 2 April. Retrieved June 10, 2025, from https://e360.yale.edu/features/anthropocene-denied

Bird, K. (2023). *The Tragedy of J. Robert Oppenheimer.* The New York Times. Retrieved June 10, 2025, from https://www.nytimes.com/2023/07/17/opinion/kai-bird-oppenheimer-christopher-nolan.html

Bjornerud, M. (2018). *Timefulness: How thinking like a geologist can help save the world.* Princeton University Press.

Bleiberg, L. (2024). *A journey 400,000 years back in human history.* BBC. Retrieved June 10, 2025, from https://www.bbc.com/travel/article/20241105-a-journey-400000-years-back-in-human-history

Bohle, M., Holzert, B., Sklair, L., & Will, F. (2025). Conclusion: The Anthropocene remade. In *The Anthropocene working group and the global debate around a new geological epoch. Emerging globalities and civilizational perspectives* (pp. 181–193). Springer. https://doi.org/10.1007/978-3-031-85175-9_6

Boivin, N., Braje, T., & Rick, T. (2024). New opportunities emerge as the Anthropocene epoch vote falls short. *Nature Ecology & Evolution, 8,* 844–845. https://doi.org/10.1038/s41559-024-02392-x

Bourzac, K. (2025). The Anthropocene is dead—long live the Anthropocene. *Engineering, 47,* 4–6. https://doi.org/10.1016/j.eng.2025.03.003

Brannen, P. (2018). *Rambling through time.* The New York Times, 27 January. Retrieved June 10, 2025, from https://www.nytimes.com/2018/01/27/opinion/rambling-through-time.html

Brzezinski, Z. (1993). *Out of control: Global turmoil on the eve of the 21st century.* Scribners.

Caballero, P., & Lõndono, P. (2022). *Redefining development: The extraordinary genesis of the sustainable development goals.* Lynne Reiner Publishers.

Cajete, G. (2000). *Native science: Natural laws of interdependence.* Clear Light Publishers.

Christian, D. (2019). "The keen longing for unified, all-embracing knowledge": Big history, cosmic evolution, and new research agendas. *Journal of Big History, III*(3), 3–18. https://doi.org/10.22339/jbh.v3i3.3320

Crutzen, P. (2002). Geology of mankind. *Nature, 415,* 23. https://doi.org/10.1038/415023a

Crutzen, P., & Stoermer, E. (2000). The 'Anthropocene'. *Global Change Newsletter, International Geosphere-Biosphere Programme, 41,* 17–18. Retrieved June 10, 2025, from https://www.mpic.de/3865097/the-anthropocene

Crutzen, P., & Stoermer, E. (2013). The Anthropocene: How can we live in a world where there is no nature without people? In L. Robin, S. Sörlin, & P. Warde (Eds.), *The future of nature: Documents of global change.* Yale University Press, Part 10. Retrieved June 10, 2025, from https://www.degruyter.com/document/doi.org/10.12987/9780300188479-041/html

Dalby, S. (2016). Framing the Anthropocene: The good, the bad and the ugly. *The Anthropocene Review, 3*(1), 33–51. https://doi.org/10.1177/2053019615618681

Damianos, A. (2024). Anthropocene angst: Authentic geology and stratigraphic sincerity. *Social Studies in Science, 55*(3). https://doi.org/10.1177/03063127241282309

Davis, W. (2009). *The Wayfinders: Why ancient wisdom matters to the modern world*. House of Anansi Press.

Edgeworth, M. (2025). Chicken bones and distorting timelines. The Guardian. Retrieved June 10, 2025, from https://www.theguardian.com/science/2025/feb/28/chicken-bones-and-distorting-timelines

Edgeworth, M., Bauer, A., Ellis, E., Finney, S., Gill, J., et al. (2024). The Anthropocene is more than a time interval. *Earth's Future, 12*(7). https://doi.org/10.1029/2024EF004831

Edgeworth, M., Ellis, E., & Gibbard, P. (2019). The chronostratigraphic method is unsuitable for determining the start of the Anthropocene. *Progress in Physical Geography, 43*(3). https://doi.org/10.1177/0309133319831673

Edgeworth, M., Richter, D., Waters, C., & Price, S. (2015). Diachronous beginnings of the Anthropocene: The lower bounding surface of anthropogenic deposits. *The Anthropocene Review, 2*(1). https://doi.org/10.1177/2053019614565394

Ellis, E. (2023). *Defining the Anthropocene*. New Scientist, 21. Retrieved June 10, 2025, from https://anthroecology.org/wp-content/uploads/2023/09/ellis_2023a.pdf

Ellis, E. (2024). *The Anthropocene event*. Geoscientist, 2 September. Retrieved June 10, 2025, from https://geoscientist.online/sections/viewpoint/the-anthropocene-event/

Erwin, D. (2024). Fall of the wild: Why pristine wilderness is a human-made myth. *Nature, 632*, 974–975. https://doi.org/10.1038/d41586-024-02761-3

Finney, S. (2014). The 'Anthropocene' as a ratified unit in the ICS International Chronostratigraphic Chart: Fundamental issues that must be addressed by the Task Group. In C. N. Waters, J. A. Zalasiewicz, M. Williams, M. Ellis, & A. M. Snelling (Eds.), *A stratigraphical basis for the Anthropocene*. Geological Society (Vol. 395, pp. 23–28). Special Publications. https://doi.org/10.1144/SP395.9

Finney, S., & Edwards, L. (2016). The "Anthropocene" epoch: Scientific decision or political statement? *GSA Today, 26*(3), 4–10. https://doi.org/10.1130/GSATG270A.1

Finney, S., & Gibbard, P. (2023). The humanities are invited to the Anthropocene Event but not to the Anthropocene Series/Epoch: A response to Chvostek (2023). *Journal of Quaternary Science, 38*(4), 461–462. https://doi.org/10.1002/jqs.3520

Gabbott, S., & Zalasiewicz, J. (2025). *Discarded: How technofossils will be our ultimate legacy*. Oxford University Press.

Ghosh, R. (2024). A fond farewell to the Anthropocene. *Issues in Science and Technology, XL*(3). Retrieved June 10, 2025, from https://issues.org/farewell-anthropocene-ghosh/

Gibbard, P., Walker, M., Bauer, A., Edgeworth, M., Edwards, L., et al. (2022). The Anthropocene as an event, not an epoch. *Journal of Quaternary Science, 37*(3), 395–399. https://doi.org/10.1002/jqs.3416

Hadly, E., & Barnosky, A. (2024). *Love it, or hate it, the Anthropocene is here to stay*. Geoscientist, 2 September. Retrieved June 10, 2025, from https://geoscientist.online/sections/viewpoint/love-it-or-hate-it-the-anthropocene-is-here-to-stay/

Hamilton, C., & Grinevald, J. (2015). Was the Anthropocene anticipated? *The Anthropocene Review, 2*(1), 59–72. https://doi.org/10.1177/2053019614567155

Henderson, E., & Vachula, R. (2024). Geologic limitations on a comprehensive Anthropocene. *Anthropocene, 46*(1). https://doi.org/10.1016/j.ancene.2024.100434

Hutton, J. (2010). *Theory of the earth (1788)*. Classic Books International.

Jackson, S. (2019). Humboldt for the Anthropocene: Humboldt's fusion of science and humanism can address contemporary challenges. *Science, 365*(6458), 1074–1076. https://doi.org/10.1126/science.aax7212

Kampmark, B. (2023). *The Oppenheimer imperative: Normalizing atomic terror*. Independent Australia, 22 August. Retrieved June 10, 2025, from. https://independentaustralia.net/life/life-display/the-oppenheimer-imperative-normalising-atomic-terror,17828

Kennedy, J. (1962). *Remarks at the America's Cup dinner, Newport, Rhode Island.* John F. Kennedy Presidential Library and Museum. Retrieved June 10, 2025, from https://www.jfklibrary.org/arc hives/other-resources/john-f-kennedy-speeches/americas-cup-dinner-19620914

Kingsbury, K., Hennigan, W., & Cohen, S. (2024). *The last survivors of the atomic bomb.* The New York Times. Retrieved June 10, 2025, from https://www.nytimes.com/interactive/2024/08/06/ opinion/hiroshima-nagasaki-atomic-bombing.html

Koch, A., Lewis, S., Brierley, C., & Maslin, M. (2024). Human impacts on the climate prior to the industrial revolution. In S.R. Brechin & S. Lee (Ed.), *Routledge handbook of climate change and society.* https://doi.org/10.4324/9781003291206-10

Koster, E. (2011). The Anthropocene: An unprecedented opportunity to advance the unique relevance of geology to societal and environmental needs. *Geoscientist, 21*(9), 18–21. https://geo scientist.online/wp-content/uploads/2025/07/Geoscientist-2011_21_09-October-1.pdf

Koster, E. (2020). Anthropocene: Transdisciplinary shorthand for human disruption of the Earth System. *Geoscience Canada, 47*(1–2), 59–64. https://doi.org/10.12789/geocanj.2020.47.160

Koster, E. (2022). Public-minded reflections from an Anthropocene Working Group meeting in Germany. *Episodes, 46*(2), 317–324. https://doi.org/10.18814/epiiugs/2022/022033

Koster, E. (2024). *Geoscience provides a needed holistic context for Earth System Governance.* Mobilizing an Earth Governance Alliance, November. Retrieved June 10, 2025, from https://earthgovernance.org/resources/geoscience-provides-a-needed-holistic-con text-for-earth-system-governance/

Koster, E. (2025a). Toward an earth-human ecosystem. *Mondial, Citizens for Global Solutions, 8*(2), 19–21. Retrieved June 10, 2025, from https://globalsolutions.org/updates/mondial-journal/tow ard-an-earth-human-ecosystem/

Koster, E. (2025b). Review of 'The Anthropocene working group and the global debate' by M. Bohle, B. Holzer, L. Sklair and F. Will. *Geology Today, 41*(4), 176. https://doi-org.ezp.lib.cam. ac.uk/10.1111/gto.12523

Koster, E., Gibbard, P., Edgeworth, M., & Jossey, R. (2022). Time is pressing: can we be good ancestors? Museum of Science, Boston Retrieved June 10, 2025, from https://www.mos.org/ blog/the-earth-around-us/good-ancestors

Koster, E., Gibbard, P., & Maslin, M. (2023). Optimizing the Anthropocene definition: an epistemological view with briefings on four conferences. *Episodes, 46*(2), 325–336. https://doi.org/ 10.18814/epiiugs/2023/023005

Koster, E. Gibbard, P., & Maslin M. (2024). The Anthropocene Event as a cultural zeitgeist in the earth-human ecosystem. *Journal of Geoethics and Social Geosciences, 1*(1), 1–41. https://doi. org/10.13127/jgsg-43

Krause, J., & Trappe, T. (2025). *Hubris: The rise, fall, and future of humanity.* Polity.

Lopez-Claros, A., Dahl, A., & Groff, M. (2020). *Global governance and the emergence of global institutions for the 21st century.* Cambridge University Press.

Lyon, H. (2017). *What's in a name?: How place names reveal our history.* Doing History in Public. Retrieved June 10, 2025, from. https://www.historyassociates.com/place-names/

MacAskill, W. (2022). *The case for longtermism.* The New York Times. Retrieved June 10, 2025, from. https://www.nytimes.com/2022/08/05/opinion/the-case-for-longtermism.html

Malhi, Y. (2017). The concept of the Anthropocene. *Annual Review of Environment and Resources, 42*, 77–104. https://doi.org/10.1146/annurev-environ-102016-060854

McCarthy, F., Head, M., Waters, C., & Zalasiewicz, J. (2025). Would adding the Anthropocene to the geologic time scale matter? *AGU Advances, 6*(2). https://doi.org/10.1029/2024AV001430

Möllers, N., Schwägerl, C. & Trischler, H. (2015). *Welcome to the Anthropocene: The Earth in our hands.* Deutsches Museum and Rachel Carson Center.

Sagan, C. (2013). *Cosmos.* Ballantyne Books.

Schwägerl, C. (2015). We aren't doomed: An interview with Paul Crutzen. In N. Möllers, C. Schwägerl, & H. Trischler (Eds.), *Welcome to the Anthropocene: The Earth in our hands* (pp. 30–36). Deutsches Museum and Rachel Carson Center.

Skelton, A., & Noone, K. (2025). The case for the Anthropocene epoch is stronger than the case for the Holocene Epoch. *Earth's Future, 13*(5). https://doi.org/10.1029/2024EF005719

Steffen, W., Broadgate, W., Deutsch, L., Gaffney, O., & Ludwig, G. (2015). The trajectory of the Anthropocene: The great acceleration. *The Anthropocene Review, 2*(1). https://doi.org/10.1177/2053019614564785

Subramanian, M. (2019). Anthropocene now: Influential panel votes to recognize Earth's new epoch. *Nature.* https://doi.org/10.1038/d41586-019-01641-5

Summerhayes, C., Zalasiewicz, J., Head, M., Syvitski, J., Barnosky, A., et al. (2024). The future extent of the Anthropocene epoch: A synthesis. *Global and Planetary Change, 242.* https://doi.org/10.1016/j.gloplacha.2024.104568

Swyngedouw, E., & Maslin, M. (2024). Responses to Ghosh (2024), The Anthropocene: Gone but not forgotten. *Issues in Science and Technology, XL*(4), Retrieved June 10, 2025. https://issues.org/farewell-anthropocene-ghosh/

The New York Times. (2011). *The Anthropocene.* The New York Times, 27 February. Retrieved June 10, 2025, from https://www.nytimes.com/2011/02/28/opinion/28mon4.html

Turner, S., & Waters, C. (2024). *Unprecedented planetary change in a human lifetime.* Geoscientist, 24 September. Retrieved June 10, 2025, from https://geoscientist.online/sections/features/unprecedented-planetary-change-in-a-human-lifetime/

UN. (2015). *Transforming our world: The 2030 agenda for sustainable development.* United Nations, Department of Economic and Social Affairs. Retrieved June 10, 2025, from https://sdgs.un.org/publications/transforming-our-world-2030-agenda-sustainable-development-17981

UN. (2023). *Halfway to 2030, world 'nowhere near' reaching global goals, UN warns.* UN News, 17 July. Retrieved June 10, 2025, from https://news.un.org/en/story/2023/07/1138777

Voosen, P. (2024). *The Anthropocene epoch is dead. Long live the Anthropocene.* Science Advisor, 5 March. https://doi.org/10.1126/science.z3wcw7b

Wagreich, M., Braun, R., & Randell, R. (2025). A new geoethics for the Anthropocene. European Geoscience Assembly, Vienna, 27 April–2 May. https://doi.org/10.5194/egusphere-egu25-17899

Walker, M., Bauer, A., Edgeworth, M., Ellis, E., Finney, S., et al. (2023). The Anthropocene is best understood as an ongoing, intensifying, diachronous event. *Boreas, 53*(1), 1–3. https://doi.org/10.1111/bor.12636

Ware, G. (2024). The Anthropocene epoch that isn't—what the decision not to label a new geological epoch means for Earth's future. The Conversation, 4 April. Retrieved June 10, 2025, from https://phys.org/news/2024-04-anthropocene-epoch-isnt-decision-geological.html

Waters, C., Syvitski, K., Galuszka, A., Hancock, G., Zalasiewicz, J., et al. (2015). Can nuclear weapons mark the beginning of the Anthropocene Epoch? *Bulletin of the Atomic Scientists, 71*(3), 46–57. Retrieved June 10, 2025, from https://thebulletin.org/2015/05/can-nuclear-weapons-fallout-mark-the-beginning-of-the-anthropocene-epoch/

Waters, C., Turner, S., Zalasiewicz, J., & Head, M. (2023). Candidate sites and other reference sections for the Global boundary Stratotype Section and Point of the Anthropocene series. *The Anthropocene Review, 10*(1), 3–24. https://doi.org/10.1177/20530196221136422

Witze, A. (2024). *It's final: The Anthropocene is not an epoch, despite protest over vote.* Nature, 20 March. https://doi.org/10.1038/d41586-024-00868-1

Zalasiewicz, J. (2008). *The world after us.* Oxford University Press.

Zalasiewicz, J., Head, M., Waters, C., Turner, S., Haff, P., et al. (2023). The Anthropocene within the geological time scale: A response to fundamental questions. *Episodes, 47*(1), 65–83. https://doi.org/10.18814/epiiugs/2023/023025

Zalasiewicz, J., Thomas, J., Waters, C., Turner, S., & Head, M. (2024). The meaning of the Anthropocene: Why it matters even without a formal geological definition. *Nature, 632,* 980–983. https://doi.org/10.1038/d41586-024-02712-y

Zalasiewicz, J., Waters, C., Williams, M. Summerhays, C., et al. (2017/18). The geological Anthropocene: Born in the Burlington House. *Geoscientist, 27*(11), 16–19.

Zerkle, A. (2024). How great was the 'Great Oxidation Event'? *Eos, 105*, 980–983. https://doi.org/10.1029/2024EO240313

Zhong, R. (2022). *For planet Earth, this might be the start of a new age.* The New York Times. Retrieved June 10, 2025, from https://www.nytimes.com/2022/12/17/climate/anthropocene-age-geology.html

Żuk, P., & Żuk, P. (2024). Beyond "geological nature," fatalistic determinism and pop-Anthropocene: Social, cultural, and political aspects of the Anthropocene. *Earth's Future, 12*(4). https://doi.org/10.1029/2023EF004045

Red Planet, Green Ethics: Navigating the Moral Terrain of Terraforming Mars

Dov Greenbaum

Abstract Terraforming Mars presents profound ethical, legal, and scientific challenges requiring structured analysis. This paper introduces exoethics, a novel framework integrating principles of planetary dignity, precautionary stewardship, distributive cosmic justice, life-centered ethics, and governance. It evaluates terraforming proposals through case studies, addressing issues such as planetary protection, biological contamination, and corporate monopolization. The paper highlights gaps in current space law, emphasizing the need for regulatory evolution and responsible planetary stewardship. As humanity stands on the threshold of extraterrestrial transformation, ethical foresight and international collaboration are essential to ensuring Mars' future aligns with the highest standards of sustainability and justice.

Keywords Exoethics · Terraforming Mars · Space law · Planetary stewardship · Space colonization

1 Introduction

The concept of terraforming Mars has captured both scientific and popular imagination, moving from theoretical discussions to proposed implementation strategies. In 2015, SpaceX CEO Elon Musk garnered widespread attention by suggesting on The Late Show that nuclear detonations could be used to warm Mars's poles, ultimately making the planet more hospitable for human habitation.[1] This proposal echoed earlier scientific discourse on planetary engineering, notably Carl Sagan's, 1973 arguments for Martian transformation (1973).

---

[1] The Late Show with Stephen Colbert (14 May 2015): Elon Musk Might Be A Super Villain. YouTube video: https://www.youtube.com/watch?v=gV6hP9wpMW8 (accessed 10 June 2025).

D. Greenbaum (✉)
Harry Radzyner Law School, Dina Recanati School of Medicine, Zvi Meitar Institute for Legal Implications of Emerging Technologies, Reichman University, Herzliya, Israel
e-mail: dov.greenbaum@runi.ac.il

Biomedical Informatics and Data Science, Yale University, New Haven, CT, USA

The prospect of terraforming Mars represents one of humanity's most ambitious potential endeavors, marking a significant leap from our limited Earth-bound environmental manipulation to planetary engineering. As space agencies and private enterprises increasingly turn their attention to Mars colonization[2] (Wattles, 2023), the ethical, legal, and social implications of deliberately altering another planet's environment demand careful consideration.

Proposals for Earth-based geoengineering have long raised ethical concerns, including environmental impact, equitable distribution of benefits and harms, moral responsibility, and the need for broad political support. Large-scale interventions, such as aerosol injection to alter solar radiation as part of solar radiation management (Pope et al., 2012) or carbon capture technologies (Wilberforce, 2021), may have unintended ecological consequences, including ocean acidification, oxygen depletion, and disruptions to planetary systems, challenging the value placed on both human and nonhuman life (Haqq-Misra, 2012).

Geoethics provides a crucial framework for evaluating the societal, political, and environmental implications of geoengineering, ensuring that such interventions are approached with caution, responsibility, and adaptability. However, geoethics is traditionally focused on Earth's geological processes; as such, it may not fully address the ethical complexities of terraforming Mars and other extra-planetary endeavors.

This chapter thus advocates for an expanded ethical framework that integrates ecological stewardship, scientific preservation, and intergenerational justice in planetary transformation.

2   Theoretical Framework: Extending Geoethics Beyond Earth

Geoethics is concerned with the ethical, social, and cultural implications of geoscience knowledge, research, practice, education, and communication. It aims to guide human behavior in relation to Earth (Peppoloni & Di Capua, 2021a, 2021b), emphasizing responsible interaction with natural environments and the proper dissemination of associated scientific knowledge (Peppoloni & Di Capua, 2022). Arguably, the core concerns of geoethics are very much Earth-bound and include managing environmental risks, protecting geodiversity as a foundation for biodiversity, and acknowledging humanity's geological force in modifying the Earth's ecosystems. It underscores the responsibility of individuals and societies to ensure sustainable and ethical stewardship of the planet (Peppoloni et al., 2019).

While some might extend this ethical responsibility to outer space, no direct geoethics equivalent exists for non-Earth settings (Chon-Torres, 2021). Earth-centric

---

² National Aeronautics and Space Administration (NASA): Mars, https://www.nasa.gov/humans-in-space/humans-to-mars/ (accessed 10 June 2025).

environmental ethics struggle to address the complexities of extraterrestrial ecosystems. These frameworks rely on assumptions grounded in Earth-specific conditions—such as the presence of biological ecosystems and longstanding human–environment relationships—that may not translate to Mars or other celestial bodies. The absence of functioning ecosystems on Mars, combined with its distinct physical and atmospheric characteristics, challenges conventional approaches to environmental preservation and restoration. This underscores the need to rethink and adapt ethical principles for planetary stewardship in extraterrestrial contexts. Existing space-related fields therefore require a more comprehensive framework for addressing the moral dimensions of human interaction with extraterrestrial landscapes.

3 Exoethics: A Structured Framework for Extraterrestrial Ethics

We propose Exoethics as a structured ethical framework for evaluating the moral, social, and cultural dimensions of human activity beyond Earth. While not a wholly new field, exoethics formalizes and operationalizes ethical insights from geoethics, environmental philosophy, space ethics, and planetary protection into a coherent, principle-based model tailored to the complexities of planetary engineering and interplanetary governance.

Rather than diverging from geoethics, exoethics extends its core commitments—planetary stewardship, ethical foresight, and geoscientific responsibility—into extraterrestrial contexts, adapting them to the unfamiliar ethical terrain of lifeless planetary environments such as Mars. This approach resonates with early calls for a more universal, or cosmocentric, ethics. For example, Marshall (1993) proposes applying terrestrial environmental principles—particularly intrinsic value and deep ecology—to space environments, arguing for their protection regardless of biological life. Similarly, Daly and Frodeman (2008) call for a reconciliation between environmental philosophy and space policy, urging that ethical frameworks not stop at Earth's atmosphere. Haqq-Misra's (2012) "ecological compass" further formalizes this shift, offering a tool for evaluating the moral consequences of planetary engineering. Together, these works form the conceptual and philosophical foundation upon which exoethics builds its structured, action-oriented framework.

What distinguishes exoethics is not the novelty of its concerns but its integration and operationalization of these diverse traditions into a unified, principle-based model designed for real-world decision-making. Unlike the broader philosophical inquiries of space ethics (Milligan & Johnson-Schwartz, 2023), astroethics (Peters, 2021), or astrobioethics (Chon-Torres, 2018), exoethics offers structured, actionable guidance for ethical engagement with planetary environments and human expansion into space.

The ethical considerations of terraforming Mars necessitate an integrative framework that builds upon established philosophical approaches of geoethics while addressing the unique challenges of extraterrestrial environments. Although the term "exoethics" synthesizes several existing ethical frameworks, it represents a necessary evolution in our moral thinking as humanity expands beyond Earth.

Exoethics is built upon five core principles that collectively offer a comprehensive framework for responsible extraterrestrial engagement. First, the concepts of **Planetary Dignity and Intrinsic Value** recognize that celestial bodies possess inherent worth beyond their utility to humans, representing unique scientific archives and aesthetic landscapes that deserve protection as part of our solar system's heritage. Second, **Precautionary Stewardship** emphasizes reversible interventions, graduated approaches to planetary modification, and placing the burden of proof on those proposing potentially harmful activities. Third, the idea of **Distributive Cosmic Justice** ensures equitable access to space resources across nations and generations, while preventing monopolization by powerful actors. Fourth, **Life-Centered Ethics** acknowledges the moral significance of potential indigenous life forms regardless of complexity, advocating for protected zones and balanced frameworks for weighing human needs against their preservation. Finally, **Governance and Accountability** principles establish transparent decision-making processes with multi-stakeholder participation, independent oversight, and adaptive legal frameworks that can evolve with technological advancement.

This framework significantly expands upon COSPAR's[3] planetary protection policies (Kminek et al., 2020), which primarily focus on preventing biological contamination. While COSPAR establishes valuable technical guidelines, it primarily recognizes scientific research value rather than intrinsic worth, takes a reactive rather than proactive approach to planetary protection, concentrates narrowly on microbial contamination rather than broader environmental and social concerns, and vests decision authority primarily within the scientific community instead of diverse stakeholders. Exoethics addresses these limitations by offering a more comprehensive ethical system that recognizes the multifaceted value of celestial bodies.

Both Marshall (1993) and Daly and Frodeman (2008) recognized that existing planetary protection frameworks, like COSPAR, are technically robust but ethically thin. Without embedding non-anthropocentric or cosmocentric ethics, space policy would lag behind the moral complexity of planetary transformation. Exoethics addresses this void by formalizing ethical protections that prioritize environmental integrity alongside scientific and technological development.

For practical application, we introduce an assessment matrix that evaluates planetary modification projects across six key criteria: reversibility, scale, life impact, resource equity, scientific value preservation, and intergenerational consent. This enables practitioners to categorize activities as presenting low, moderate, or high ethical concerns and adjust their approach accordingly. The framework also establishes a decision protocol requiring comprehensive assessment against all principles, consultation with diverse stakeholders (including scientific, policy, ethical, indigenous, and future generation perspectives), analysis of less invasive alternatives, development of harm mitigation strategies, and implementation of ongoing monitoring systems.

---

[3] Committee on Space Research (COSPAR).

Ultimately, exoethics offers what Space Ethics currently lacks: a principled and practical framework, concrete tools for ethical decision-making, an explicit focus on justice, aesthetics, governance, and legacy, and a philosophical clarity that is grounded in but not limited to Earth-based models—marking the next step in moral reasoning for space, one that draws from geoethics while transcending its terrestrial constraints and moving beyond the theoretical diffusion of existing space ethics discourse.

By providing this structured approach, exoethics aims to offer a practical and adaptable framework for navigating the complex moral terrain of extraterrestrial activities. Recognizing that ethical, legal, and technological landscapes are constantly evolving, this framework remains dynamic—integrating new insights, adapting to emerging challenges, and discarding outdated principles when necessary. Its agility ensures that humanity's expansion into space reflects our highest ethical aspirations rather than repeating historical patterns of exploitation and environmental degradation.

4   Colonizing Mars

Carl Sagan (1994) argued that planetary civilizations face existential threats and that humanity must become spacefaring for survival. This perspective supports Mars colonization as an insurance policy against Earth-bound catastrophes (Valentine, 2012). Beyond survival, colonization advances science and technology (Szocik et al., 2020), fulfilling humanity's drive for exploration (Sagan, 1991). A self-sustaining Martian settlement could accelerate innovation in engineering, space agriculture, and life support systems, benefiting both Mars and Earth. Mars also offers unique scientific insights into planetary evolution, climate change, and the search for extraterrestrial life (Szocik et al., 2020).

Economically, Mars has resource extraction potential, but long-term viability remains uncertain (Bruhns & Haqq-Misra, 2016). Colonization is often framed as a step toward a multi-planetary future (García Bonilla, 2021). However, significant ethical, legal, and logistical challenges persist. Critics warn of biological contamination, which could disrupt potential Martian ecosystems (Chon-Torres, 2021), and argue that resources should be prioritized for Earth's climate crisis (Atri et al., 2024; Schwartz, 2019). Legal uncertainties include property rights, as the Outer Space Treaty (OST) prohibits national sovereignty over celestial bodies. The harsh Martian environment, with its thin atmosphere, radiation, and water scarcity, further complicates long-term habitation.

Mars colonization also risks reinforcing global inequalities, benefiting technologically advanced nations and private corporations while diverting resources from urgent Earth-based needs (McKaig et al., 2024; Stockwell, 2020). Additionally, irreversible planetary changes could obscure Mars' past, preventing future generations from determining whether it once harbored life. These challenges fuel ongoing debates about the feasibility and ethics of Martian settlement.

While Mars offers opportunities for scientific discovery and human expansion, sustaining life remains a major hurdle. Permanent habitation may require large-scale environmental modification, as Mars' current conditions make long-term survival impossible without terraforming—the transformation of Mars to resemble Earth's environment.

5 Core Challenges in Terraforming Mars

Terraforming is the theoretical process of altering a planet's atmosphere, temperature, surface, or ecology to make it more Earth-like and capable of supporting human and other terrestrial life (Iliopoulos & Esteban, 2020). This concept involves large-scale planetary engineering aimed at creating a self-sustaining biosphere, eliminating the need for artificial life support systems (Fogg, 1995). Originally a science fiction idea, the term was popularized by writer Jack Williamson (1942) and is now primarily associated with transforming Mars or other celestial bodies into habitable environments (Priya & Ansari, 2024).

In contrast, geoengineering refers to deliberate, large-scale interventions in Earth's climate system to mitigate or counteract climate change. While both terraforming and geoengineering involve planetary-scale modifications, their objectives and ethical considerations differ (Haqq-Misra, 2012). Terraforming aims to create a self-sustaining biosphere on another planet, while geoengineering modifies Earth's environment to address climate challenges like temperature regulation and carbon sequestration (Prunariu & Tulbure, 2021). Whereas geoengineering is primarily concerned with reducing global warming, terraforming seeks to accomplish the opposite, the warming of Mars.

5.1 Scientific and Technical Challenges

Terraforming Mars remains speculative (Dodge, 2022) and presents major technical and scientific challenges. Its thin atmosphere, only 1% of Earth's and composed mainly of carbon dioxide (Zubrin & McKay, 1993), makes creating sufficient atmospheric pressure difficult. Mars' average temperature of $-67\,°C$ necessitates temperature regulation through solar reflectors, greenhouse gas emissions, or asteroid bombardment, all requiring immense energy (Priya & Ansari, 2024). The planet's lack of a global magnetosphere exposes settlers to harmful radiation, with potential solutions including artificial magnetic shields or underground habitats (Pulsiri et al., 2022).

Water availability remains a challenge, as converting frozen polar water to liquid form demands significant energy. Proposed solutions involve solar reflectors and surface modifications to increase heat absorption (Hempsell et al., 2018). Establishing a biosphere requires creating self-sustaining carbon and oxygen cycles, likely through photosynthetic organisms, though their survival in Mars' extreme conditions is uncertain. Current engineering capabilities are insufficient for planetary-scale transformation (Iliopoulos & Esteban, 2020), and long-term success would require sustained international collaboration over decades (Szocik et al., 2020).

Additional sustainability concerns include atmospheric retention and resource extraction (Cypser, 1993), as Mars' harsh conditions pose obstacles to creating a human-compatible environment (Lopez et al., 2019). While theoretical planetary modification methods exist, they remain beyond technological feasibility and require macro-engineering breakthroughs (Hempsell et al., 2018). Resource extraction further complicates governance (Gnanesh, 2021), and preventing corporate monopolization is essential for equitable access. Extraterrestrial resource use currently lacks robust legal and ethical frameworks, necessitating international agreements to balance economic interests with planetary protection (Tiwari, 2021; Valente et al., 2025).

Beyond technological hurdles, making Mars habitable requires addressing legal and ethical challenges of planetary stewardship. Success depends on innovative engineering, strong international governance, and preserving Mars as a shared scientific and ecological resource rather than an unregulated frontier. Terraforming also raises critical questions about environmental responsibility, planetary protection, and the balance between exploration and preservation.

5.2 Core Ethical Challenges Through an Exoethics Lens

Through the lens of the exoethics framework, terraforming Mars presents complex ethical challenges that require systematic analysis. Each core principle of exoethics draws from ethical traditions across virtue ethics, deontological reasoning, consequentialism, and theories of justice. For example, Planetary Dignity echoes non-anthropocentric environmental ethics, while Distributive Cosmic Justice adapts concepts of equity across space and generations. Precautionary Stewardship reflects both virtue-based restraint and the precautionary principle seen in Earth-based geoethics. Each core principle of exoethics highlights different aspects of the moral landscape that must be addressed.

Planetary Dignity and Intrinsic Value question humanity's moral authority to permanently alter celestial bodies (Peters, 2015). Mars serves as both a scientific archive and a potential biological frontier, where forward contamination from Earth could disrupt ecosystems and compromise irreplaceable discoveries (de Zwart et al., 2021). Some argue that Mars has already exchanged biological material with Earth via meteorites (Beech et al., 2018; Kawaguchi et al., 2020), yet exoethics advocates for conservation when irreversible decisions affect environments of scientific and biological significance.

Precautionary Stewardship emphasizes the risks of irreversible terraforming. If microbial life exists on Mars, large-scale modifications could trigger extinction events. Exoethics supports a graduated intervention approach—prioritizing scientific exploration and limited, reversible experiments before full-scale planetary transformation. This allows for careful evaluation while ensuring that the burden of proof remains with those advocating for irreversible change (Chon-Torres, 2021).

Distributive Cosmic Justice raises concerns about resource allocation. Committing vast resources to Mars, while Earth faces urgent environmental challenges prompts questions of equity. Critics warn that massive investments in terraforming may primarily benefit wealthy nations or corporations, exacerbating existing inequalities rather than resolving them (Profitiliotis & Loizidou, 2019). Exoethics demands mechanisms to ensure that benefits are fairly distributed across both time and humanity, preventing monopolization by early actors.

Life-Centered Ethics weighs human expansion against the protection of potential Martian life. While some argue that transforming Mars for human habitation outweighs preserving microbial life, exoethics advocates for the establishment of protected zones, akin to terrestrial biosphere reserves. This approach balances human aspirations with the moral significance of indigenous Martian microorganisms, promoting coexistence through planetary zoning and protection frameworks.

Governance and Accountability address intergenerational ethics. Terraforming would obligate future generations to maintain planetary-scale interventions they did not consent to, potentially imposing unforeseen ecological and societal burdens (Peppoloni & Di Capua, 2021a, 2021b). While history shows that technological advancements often shape environments without future consent, exoethics emphasizes transparent decision-making, broad stakeholder inclusion, and formal representation of future generations' interests to safeguard long-term options.

The practical application of exoethics to Mars terraforming involves evaluating projects across six key criteria: **reversibility, scale, life impact, resource equity, scientific value preservation, and intergenerational consent**. Most terraforming proposals would trigger high ethical concerns in multiple areas, underscoring the need for modified approaches, robust international oversight, and a phased implementation strategy that prioritizes limited, reversible interventions in designated zones.

Applying exoethics shifts the debate from polarization to balanced, responsible planetary stewardship. This framework offers both a philosophical foundation and practical tools to ensure that planetary engineering reflects humanity's highest ethical aspirations rather than repeating historical patterns of exploitation and environmental degradation.

5.3 Dual Use Technologies

Many space technologies have dual-use potential for civilian and military applications (Melamed et al., 2024). Geoengineering has long drawn military interest (Hamilton, 2014), with past weaponization, such as cloud seeding in the Vietnam War (McGee et al., 2020). While the 1977 Environmental Modification Convention (ENMOD) bans military use, 101 nations, including some spacefaring ones, have not signed it, and its applicability to extraterrestrial bodies remains unclear (UN, 1977).

Terraforming technologies—atmospheric modification, genetic engineering, and planetary engineering—pose similar risks if unregulated. Designed for habitability, they could be repurposed for strategic dominance, with planetary weather control enabling ecological disruption or territorial claims (Jarose, 2024). Exoenvironmental Warfare, where nations or corporations manipulate Martian resources like water ice or rare metals, could escalate conflicts. Weaponized atmospheric modifications might deny breathable air, while bioengineered microorganisms could disrupt rival colonies' infrastructure, food, or health.

On Earth, geoengineering raises ethical concerns due to unintended consequences and political misuse (Young, 2023). Terraforming presents an even greater risk, where monopolization of planetary engineering could enable climate warfare. Historical failures in enforcement highlight the challenge of preventing militarization (Melamed et al., 2024; Young, 2023). Instead of abandoning terraforming, international safeguards, treaties, and monitoring systems must be implemented. Properly regulated, planetary engineering could also benefit Earth by mitigating climate change, advancing clean energy, and strengthening biosphere resilience.

6 Legal and Governance Challenges of Terraforming Mars

Terraforming Mars presents significant legal, ethical, and governance challenges, particularly concerning planetary protection, international law, and public–private conflicts. A primary concern is harmful contamination, as introducing Earth-based organisms could violate the Outer Space Treaty (OST) (UN, 1967, Article IX), which requires states to prevent contamination of celestial bodies (Altabef, 2021). While COSPAR planetary protection protocols (COSPAR, 2024) and the Moon Agreement (UN, 1984, Article 7.1) further discourage altering a celestial body's identity, they lack enforcement mechanisms, necessitating new legal frameworks to regulate planetary-scale modifications (Chan, 2021).

Another major issue is the non-appropriation principle (UN, 1967, Article II), which prohibits states and private entities from claiming celestial bodies (Bruhns & Haqq-Misra, 2016). However, the Artemis Accords argue that resource extraction does not constitute sovereignty, aligning with space policies of the US, Luxembourg, and the UAE (ud Din, 2022). This ambiguity raises concerns over monopolization and unequal access to Martian resources (Lucas-Rhimbassen, 2021). In the absence of governance, private entities may prioritize profit at the expense of sustainability, exacerbating global inequalities (Nugraha, 2022). Some scholars propose granting legal personhood to celestial bodies as additional safeguards against reckless modifications (Altabef, 2021; Siebrits, 2021).

Calls for OST revisions (Deplano, 2021) include proposals for a "Mars Agreement," an extension of existing space treaties to regulate long-term planetary development (Torres-Spelliscy, 2024).

The involvement of private companies in terraforming introduces further challenges in accountability, oversight, and planetary protection. While government-led missions prioritize planetary ethics, profit-driven private firms may favor short-term economic returns over sustainability (Dey & Jagadanandan, 2025). The OST does not adequately regulate commercial activities, increasing risks of monopolization and environmental neglect. However, private-sector innovation could accelerate resource utilization and technology development, making regulatory oversight essential (Lucas-Rhimbassen, 2021).

7 Exoethics in Practice: Case Studies of Martian Terraforming Proposals

To illustrate the practical application of the exoethics framework, we present three hypothetical scenarios involving different approaches to Martian transformation. Each case study demonstrates how the five core principles of exoethics can guide ethical and legal decision-making in specific contexts of Mars colonization.

7.1 Case Study 1: Localized Atmospheric Modification Dome

Proposal: A private aerospace corporation proposes establishing a 50 km-diameter atmospheric modification dome over Valles Marineris, home to glaciers which may host Martian life. The dome would create Earth-like atmospheric pressure and composition within its boundaries using solar-powered compressors and chemical processors that extract gases from Martian soil. The corporation argues this approach allows for habitation while minimizing planetary-scale alterations.

Exoethics Assessment: The proposal acknowledges Mars' intrinsic value by limiting modifications to a specific region rather than undertaking global transformation. However, Valles Marineris, a unique geological formation with significant scientific importance and potential for life, would face irreversible alteration. While the contained nature of the dome allows for a phased approach, enabling limited testing before broader implementation, decommissioning remains complex and would leave lasting traces of human activity. The corporation's exclusive rights to the dome and surrounding territory raise concerns about monopolization and restricted access. Although prior robotic missions have not detected microbial life, subsurface investigations remain incomplete. Governance is another concern, as a private corporation retains operational control, with no mechanisms for public input or transparent decision-making on future expansions or environmental impact. The proposal also raises questions about compliance with the OST's Article IX requiring avoidance of "harmful contamination" of celestial bodies. Without clear definitions of what constitutes "harmful," the atmospheric and geological modifications could be contested under existing law.

Recommended Modifications: The exoethics framework would suggest relocating the dome to an area of less scientific significance, implementing a more equitable access model with international participation, establishing robust life-detection protocols with predefined conservation measures, and creating an inclusive governance structure with meaningful public participation. From a legal perspective, the exclusive territorial claim should be replaced with a limited-term usage license approved by an international body, with clear provisions for benefit sharing and eventual transfer to international management.

7.2 Case Study 2: Orbital Solar Mirrors for Polar Warming

Proposal: An international consortium of national space agencies proposes deploying large orbital solar reflectors to concentrate sunlight on Mars' polar ice caps, gradually releasing CO_2 and water vapor to thicken the atmosphere and initiate a greenhouse effect. The consortium emphasizes this represents a "light touch" approach that accelerates natural processes without introducing new materials to the Martian environment.

Exoethics Assessment: The proposal enhances Mars' natural warming without introducing artificial chemicals but risks permanently altering polar regions, which hold valuable climate records. Implementing it gradually allows for ongoing assessment and potential decommissioning if issues arise, with safety addressed through climate modeling. While the international consortium ensures broad participation, access rights to a transformed Mars remain unclear. Though unlikely to host active life, melting polar ice could activate dormant microorganisms, necessitating biological monitoring and protected zones. Governance would fall under an expanded COSPAR framework with global representation, public hearings, and phased approvals. Legally, the plan faces scrutiny under the Outer Space Treaty's contamination clause and the Moon Treaty's prohibition on disrupting celestial environments. Liability concerns and jurisdictional complexities further complicate regulation, as existing space law lacks clear mechanisms for addressing ecological damage to celestial bodies.

Recommended Modifications: The exoethics framework would suggest first conducting extensive in situ research of polar regions to preserve scientific data, establishing binding agreements on equitable access rights to a transformed Mars before project initiation, and creating a more robust intergenerational representation mechanism through a dedicated "Future Generations Council" with veto power over irreversible decisions. From a legal perspective, the consortium should seek formal authorization through UN COPUOS and even consider establishing a specific multilateral treaty defining permissible climate modification parameters and liability protocols before deployment begins.

7.3   Case Study 3: Bioengineered Extremophile Deployment

Proposal: A university-led research collaboration proposes introducing bioengineered extremophile microorganisms to the Martian surface. These organisms, derived from Earth's most resilient bacteria but genetically modified for Martian conditions, would gradually transform Martian regolith, release oxygen, and create conditions more amenable to subsequent Earth life. The researchers argue this "biological terraforming" mimics Earth's own evolutionary history and represents the most natural approach to planetary transformation.

Exoethics Assessment: This proposal poses significant ethical and legal challenges, as introducing engineered life would permanently alter Mars' geochemistry and potentially outcompete native microorganisms, if they exist. The irreversibility of bioengineered organisms, potentially without competing natural predators, raises concerns about long-term consequences, with no clear mechanism for recall or control. Legally, this proposal directly violates COSPAR's Planetary Protection Policy and the Outer Space Treaty's ban on harmful contamination. It also raises complex biosafety and biosecurity issues, with no extraterrestrial application of the Cartagena Protocol on Biosafety. Additionally, liability remains unresolved, as assessing damage from evolving organisms would be nearly impossible.

Recommended Modifications: The exoethics framework would likely suggest rejecting this proposal in its current form due to multiple ethical and legal concerns. Modified approaches could include: Restricting deployment to fully contained experimental habitats rather than open release; requiring decades of preliminary in situ research to conclusively determine the absence of indigenous life before any consideration of wider deployment; developing robust containment and control mechanisms; and establishing an internationally binding treaty specifically governing biological planetary modification. From a legal perspective, a new protocol specifically addressing extraterrestrial biosafety would need to be developed and widely ratified before any version of this proposal could be reconsidered.

These case studies highlight exoethics as a structured method for evaluating terraforming proposals, ensuring ethical, legal, and scientific integrity. Current space law lacks specific guidance for planetary engineering, necessitating specialized frameworks that integrate ethical principles with clear legal standards. Responsible planetary transformation requires foresight to avoid repeating Earth's history of exploitation and environmental harm, ensuring Mars' future aligns with humanity's highest ethical aspirations.

8   Conclusion: Navigating the Ethical Frontier

"The day we have the power to terraform Mars, to turn Mars into Earth, we will have the power to turn Earth into Earth."[4] This underscores that terraforming is not just a technological milestone but a profound ethical test. As we stand on the brink of planetary transformation, the key question is not just whether we can terraform Mars, but whether we should—and under what principles and constraints. The exoethics framework provides a structured approach to these challenges, balancing Planetary Dignity, Precautionary Stewardship, Distributive Cosmic Justice, Life-Centered Ethics, and Governance and Accountability. By systematically evaluating terraforming proposals, exoethics identifies ethical concerns and proposes modifications to align with responsible cosmic stewardship. The ambition of exoethics is not to replace or bypass prior traditions but to offer a structured, decision-ready model—bridging environmental ethics, space policy, and geoethical responsibility for applications in planetary engineering.

Moving forward, existing legal frameworks, particularly the Outer Space Treaty, must evolve to govern planetary engineering effectively. With private-sector expansion, oversight must balance innovation with ethics. Given our limited understanding of Mars, a gradual, reversible approach—beginning with scientific study and small-scale interventions—remains the most ethical path. Key steps include forming interdisciplinary working groups within UN COPUOS to establish legal frameworks, designating protected scientific zones to preserve Mars' unique value, and developing transparent international protocols for evaluating terraforming proposals. Investing in minimally invasive, reversible technologies and ensuring global participation in decision-making are also critical.

Terraforming decisions will set precedents for humanity's relationship with other celestial bodies. If Martian life exists, the ethical stakes are even higher, forcing us to define our responsibilities toward non-terrestrial organisms. Even if Mars is lifeless, it remains an irreplaceable scientific archive and potential future home, requiring careful stewardship. Terraforming, if pursued, must prioritize ethics over economic or national interests, ensuring equity, environmental protection, and adaptability for future generations.

Ultimately, our approach to Mars defines what kind of cosmic species we wish to become. Will we repeat Earth's history of exploitation and environmental destruction, or will we pioneer a future of sustainable stewardship? Exoethics offers a path toward responsible planetary transformation, ensuring Mars becomes a source of pride, not regret, for future generations. There is time to get this right—to develop the ethical, legal, and scientific structures necessary for a future where our expansion into space reflects our highest aspirations rather than our lowest impulses.

---

[4] Terraforming Mars with Neil deGrasse Tyson, 7 May 2018. YouTube video: https://www.youtube.com/watch?v=7XdkKMhAdnA (accessed 10 June 2025).

References

Altabef, W. B. (2021). The legal man in the moon: Exploring environmental personhood for celestial bodies. *Chicago Journal of International Law, 21*(2), 476. Retrieved June 10, 2025, from https://chicagounbound.uchicago.edu/cjil/vol21/iss2/7

Athar ud Din. (2022). The Artemis accords: The end of multilateralism in the management of outer space? *Astropolitics, 20*(2–3), 135–150. https://doi.org/10.1080/14777622.2022.2144241

Atri, D., Umansky, P., & Sreenivasan, K. R. (2024). Sustainability as a core principle of space and planetary exploration. *Space Policy, 70*, Article 101636. https://doi.org/10.1016/j.spacepol.2024.101636

Beech, M., Coulson, I. M., & Comte, M. (2018). Lithopanspermia—The terrestrial input during the past 550 million years. *American Journal of Astronomy and Astrophysics, 6*(3), 81–90. https://doi.org/10.11648/j.ajaa.20180603.14

Bruhns, S., & Haqq-Misra, J. (2016). A pragmatic approach to sovereignty on Mars. *Space Policy, 38*, 57–63. https://doi.org/10.1016/j.spacepol.2016.05.008

Chan, Y. C. (2021). Protecting the million-year picnic: The importance of importing the rule of law to Mars. In: *Assessing a mars agreement including human settlements* (pp. 181–191). Springer.

Chon-Torres, O. A. (2018). Astrobioethics. *International Journal of Astrobiology, 17*(1), 51–56. https://doi.org/10.1017/S1473550417000064

Chon-Torres, O. A. (2021). Astrobioethics: Epistemological, astrotheological, and interplanetary issues. *Astrobiology: Science, Ethics, and Public Policy*, 1–15. https://doi.org/10.1002/9781119711186.ch1

COSPAR. (2024). COSPAR policy on planetary protection. *Space Research Today, 220*, 10–34.

Cypser, D. A. (1993). International law and policy of extraterrestrial planetary protection. *Jurimetrics, 33*(2), 315–339.

Daly, E. M., & Frodeman, R. (2008). Separated at birth, signs of rapprochement: Environmental ethics and space exploration. *Ethics and the Environment, 13*(1), 135–151. https://muse.jhu.edu/article/239385

Deplano, R. (2021). The Artemis accords: Evolution or revolution in international space law? *International & Comparative Law Quarterly, 70*(3), 799–819. https://doi.org/10.1017/S0020589321000142

Dey, A., & Jagadanandan, J. (2025). Balancing commercialisation and sustainability in outer space: Addressing new challenges. *Acta Astronautica, 229*, 895–900. https://doi.org/10.1016/j.actaastro.2025.02.010

de Zwart, M., Henderson, S., & Neef, R. (2021). The principle of "Harmful Contamination" applied to human missions to Mars. *Journal of Space Law, 45*(2), 276–318. Retrieved June 10, 2025, from https://airandspacelaw.olemiss.edu/wp-content/uploads/2023/08/45.2-1.pdf

Dodge, M. (2022). Celestial agriculture: Law & policy governing the use of in situ resources for space settlements. *North Dakota Law Review, 97*, 161. Retrieved June 10, 2025, from: https://law.und.edu/_files/docs/ndlr/pdf/issues/97/2/97ndlr161.pdf

Fogg, M. J. (1995). *Terraforming: Engineering planetary environments. Warrendale.*

García Bonilla, J. (2021). How five fundamental human rights could be violated in privately-funded space settlements and the role of the Mars agreement in their protection. In A. Froehlich (Ed.), *Assessing a Mars agreement including human settlements* (pp. 37–49). Springer. https://doi.org/10.1007/978-3-030-65013-1

Gnanesh, R. S. (2021). A tale of two planets in international space law: Limitations to the freedom of exploration and use. In A. Froehlich (Ed.), *Assessing a Mars agreement including human settlements* (pp. 167–180). Springer. https://doi.org/10.1007/978-3-030-65013-1

Hamilton, C. (2014). Geoengineering and the politics of science. *Bulletin of the Atomic Scientists, 70*(3), 17–26. https://doi.org/10.1177/0096340214531173

Haqq-Misra, J. (2012). An ecological compass for planetary engineering. *Astrobiology, 12*(10), 985–997. https://doi.org/10.1089/ast.2011.0796

Hempsell, M., Longstaff, R., & Alexandra, S. (2018). Preserving geostationary orbit: The next steps. *Journal of the British Interplanetary Society, 71*, 314–322. Retrieved June 10, 2025, from https://www.bis-space.com/membership/jbis/2018/JBIS-v71-no09-September-2018-dpe452.pdf

Iliopoulos, N., & Esteban, M. (2020). Sustainable space exploration and its relevance to the privatization of space ventures. *Acta Astronautica, 167*, 85–92. https://doi.org/10.1016/j.actaastro.2019.09.037

Jarose, J. (2024). A sleeping giant? The ENMOD convention as a limit on intentional environmental harm in armed conflict and beyond. *American Journal of International Law, 118*(3), 468–511. https://doi.org/10.1017/ajil.2024.15

Kawaguchi, Y., Shibuya, M., Kinoshita, I., Yatabe, J., Narumi, I., et al. (2020). DNA damage and survival time course of deinococcal cell pellets during 3 years of exposure to outer space. *Frontiers in Microbiology, 11*, 2050. https://doi.org/10.3389/fmicb.2020.02050

Kminek, G., Hedman, N., Ammannito, E., Deshevaya, E., Doran, P., et al. (2020). COSPAR policy on planetary protection. *Space Research Today, 208*, 10–22. https://doi.org/10.1016/j.srt.2020.07.009

Lopez, J. V., Peixoto, R. S., & Rosado, A. S. (2019). Inevitable future: Space colonization beyond Earth with microbes first. *FEMS Microbiology Ecology, 95*(10), fiz127. https://doi.org/10.1093/femsec/fiz127

Lucas-Rhimbassen, M. (2021). On the province of all Marskind. In A. Froehlich (Ed.), *Assessing a Mars agreement including human settlements* (pp. 27–35). Springer. https://doi.org/10.1007/978-3-030-65013-1

Marshall, A. (1993). Ethics and the extraterrestrial environment. *Journal of Applied Philosophy, 10*(2), 227–236. https://doi.org/10.1111/j.1468-5930.1993.tb00078.x

McGee, J., Brent, K., McDonald, J., & Heyward, C. (2020). International governance of solar radiation management: Does the ENMOD convention deserve a closer look? *Carbon & Climate Law Review, 14*(4), 294–305. Retrieved June 10, 2025, from https://ssrn.com/abstract=3806914

McKaig, J., Caro, T., Burton, D., Tavares, F., & Vidaurri, M. (2024). Chapter 10: Planetary protection—History, science, and the future. *Astrobiology, 24*(S1). https://doi.org/10.1089/ast.2021.0112

Melamed, A., Rao, A., de Rohan Willner, O., & Kreps, S. (2024). Going to outer space with new space: The rise and consequences of evolving public-private partnerships. *Space Policy, 68*, Article 101626. https://doi.org/10.1016/j.spacepol.2023.101626

Milligan, T., & Johnson-Schwartz, J. S. (2023). Space ethics. In *The Routledge handbook of social studies of outer space* (pp. 108–120). Routledge.

Nugraha, T. R. (2022). The prospect of interplanetary mission: Are we ready? *Padjadjaran Journal of International Law, 6*(1), 60–75. https://doi.org/10.23920/pjil.v6i1.821

Peppoloni, S., & Di Capua, G. (2021). Geoethics as global ethics to face grand challenges for humanity. In G. Di Capua, P.T. Bobrowsky, S.W. Kieffer, & C. Palinkas (Eds.), *Geoethics: Status and future perspectives*. Geological society (Vol. 508, pp. 13–29). Special Publications. https://doi.org/10.1144/SP508-2020-146

Peppoloni, S., & Di Capua, G. (2021). Current definition and vision of geoethics. In M. Bohle & E. Marone (Eds.), *Geo-societal narratives: Contextualising geosciences* (pp. 17–28). Palgrave Macmillan. https://doi.org/10.1007/978-3-030-79028-8_2.

Peppoloni, S., & Di Capua, G. (2022). *Manifesto for an ethics of responsibility towards the earth*. Springer. https://doi.org/10.1007/978-3-030-98044-3

Peppoloni, S., Bilham, N., & Di Capua, G. (2019). Contemporary geoethics within the geosciences. In M. Bohle (Ed.), *Exploring geoethics: Ethical implications, societal contexts, and professional obligations of the geosciences* (pp. 25–70). Springer. https://doi.org/10.1007/978-3-030-12010-8_2

Peters, T. (2015). Ten ethical issues in exploring our Solar Ghetto. *Journal of Astrobiology & Outreach, 4*(1). https://doi.org/10.4172/2332-2519.1000149

Peters, T. (2021). Astroethics for earthlings: our responsibility to the galactic commons. In O.A. Chon Torres, T. Peters, J. Seckbach, & R. Gordon (Eds.), *Astrobiology: Science, ethics, and public policy* (pp. 17–56). Scrivener Publishing LLC. https://doi.org/10.1002/9781119711186.ch2

Pope, F. D., Braesicke, P., Grainger, R. G., Kalberer, M., Watson, I. M., et al. (2012). Stratospheric aerosol particles and solar-radiation management. *Nature Climate Change, 2*(10), 713–719. https://doi.org/10.1038/nclimate1528

Priya, T. R., & Ansari, S. A. (2024). Terraforming: Modifying a planet into planetary habitability that's similar to Earth. *Agriculture and Food: E-Newsletter, 6*(8), 302–305. Retrieved June 19, 2025, from https://www.researchgate.net/publication/383571284_Terraforming_Modifying_a_Planet_into_Planetary_Habitability_that's_Similar_to_the_Earth

Profitiliotis, G., & Loizidou, M. (2019). Planetary protection issues of private endeavors in research, exploration, and human access to space: An environmental economics approach to backward contamination. *Space Policy, 50.* https://doi.org/10.1016/j.spacepol.2019.101332

Prunariu, D. D., & Tulbure, I. (2021). Chances and challenges of terraforming other planets. *International Multidisciplinary Scientific GeoConference: SGEM, 21*(6.1), 339–346.

Pulsiri, N., Proctor, D., Cathcart, R. B., & Buteler, J. O. (2022). Mars terraforming: A new plan for the red planet. In *2022 Portland International Conference on Management of Engineering and Technology (PICMET)* (pp. 1–5). https://doi.org/10.23919/PICMET53225.2022.9882835

Sagan, C. (1973). Planetary engineering on Mars. *Icarus, 20*(4), 513–514. https://doi.org/10.1016/0019-1035(73)90026-2

Sagan, C. (1991). Why send humans to Mars? *Issues in Science and Technology, 7*(3), 80–85.

Sagan, C. (1994). *Pale blue dot: A vision of the human future in space.* Random House.

Schwartz, J. S. J., & Milligan, T. (Eds.) (2016) *The ethics of space exploration.* Springer. https://doi.org/10.1007/978-3-319-39827-3

Schwartz, J. S. J. (2019). Where no planetary protection policy has gone before. *International Journal of Astrobiology, 18*(4), 353–361. https://doi.org/10.1017/S1473550418000228

Siebrits, A. (2021). Who speaks for Mars? The responsibility to protect and the search for life. In A. Froehlich (Ed.), *Assessing a Mars agreement including human settlements* (pp. 113–127). Springer. https://doi.org/10.1007/978-3-030-65013-1

Stockwell, S. (2020). Legal 'Black Holes' in outer space: The regulation of private space companies. *E-International Relations*, July 20. Retrieved June 10, 2025, from https://www.e-ir.info/pdf/86284

Szocik, K., Abood, S., Impey, C., Shelhamer, M., Haqq-Misra, J., et al. (2020). Visions of a Martian future. *Futures, 117.* https://doi.org/10.1016/j.futures.2019.102514

Tiwari, S. (2021). Factors influencing the future Martian population. In A. Froehlich (Ed.), *Assessing a Mars agreement including human settlements.* Studies in space policy (Vol. 30, pp. 85–98). Springer. https://doi.org/10.1007/978-3-030-65013-1_8

Torres-Spelliscy, G. (2024). Settling new frontiers: Human colonies beyond earth and the legal regulations governing their establishment. Available at SSRN 4860023.

UN. (1967). *Treaty on principles governing the activities of states in the exploration and use of outer space, including the Moon and other celestial bodies.* United Nations. Retrieved June 10, 2025, from https://www.unoosa.org/oosa/en/ourwork/spacelaw/treaties/introouterspacetreaty.html

UN. (1977). *Convention on the prohibition of military or any other hostile use of environmental modification techniques (ENMOD).* United Nations, Geneva. Retrieved June 10, 2025, from https://disarmament.unoda.org/enmod/

UN. (1984). *Agreement governing the activities of states on the Moon and other celestial bodies.* United Nations. Retrieved June 10, 2025, from https://www.unoosa.org/oosa/en/ourwork/spacelaw/treaties/intromoon-agreement.html

Valentine, D. (2012). Exit strategy: Profit, cosmology, and the future of humans in space. *Anthropological Quarterly, 85*(4), 1045–1067. https://doi.org/10.1353/anq.2012.0073

Valente, M., Caviggioli, F., & Agostini, L. (2025). Space economy and sustainability: A systematic review. *Sustainable Development.* https://doi.org/10.1002/sd.3383

Wattles, J. (2023). *Colonizing Mars could be dangerous and ridiculously expensive. Elon Musk wants to do it anyway.* CNN Business. Retrieved June 10, 2025, from, https://www.cnn.com/2020/09/08/tech/spacex-mars-profit-scn/index.html

Williamson, J. (1942). Collision orbit. *Astounding Science Fiction, 29*(6), 9–38.

Young, D. N. (2023). Considering stratospheric aerosol injections beyond an environmental frame: The intelligible 'emergency' techno-fix and preemptive security. *European Journal of International Security, 8*(2), 262–280. https://doi.org/10.1017/eis.2023.4

Wilberforce, T., Olabi, A. G., Sayed, E. T., Elsaid, K., & Abdelkareem, M. A. (2021). Progress in carbon capture technologies. *Science of the Total Environment, 761*, 143203.

Geoscience Diplomacy at the Edge of the Continental Shelf

Munira Raji and Larry Awosika

Abstract Geoscience diplomacy bridges geoscientific expertise and international policymaking to address complex global challenges. A great example is the Commission on the Limits of the Continental Shelf (CLCS), established under the United Nations Convention on the Law of the Sea (UNCLOS). This Commission integrates geoscientific data and diplomatic negotiation to support coastal states in extending their continental shelves beyond 200 nautical miles. Through the implementation of Article 76, the CLCS facilitates technical assessments, promotes cooperation and provides peaceful mechanisms for resolving maritime boundary disputes. Beyond legal delineation, geoscience diplomacy is important in maritime governance, transboundary water management, energy security, disaster risk reduction, and climate change adaptation. As pressures on shared resources grow, international collaboration in geosciences becomes increasingly important to ensure sustainability and equity. This paper explores how geoscience diplomacy is applied in real cases, such as Nigeria's successful continental shelf extension submission to the CLCS. It highlights how geoscience-informed diplomacy contributes to peaceful and sustainable governance of maritime and geological resources.

Keywords Geoscience diplomacy · United Nations · Continental shelves · UNCLOS · International collaboration

M. Raji (✉)
Sustainable Earth Institute, University of Plymouth, Drake Circus, Plymouth, Devon, UK
e-mail: munira.raji@plymouth.ac.uk

L. Awosika
Commission on the Limits of the Continental Shelf (CLCS), United Nations, NY, USA

Nigeria Extended Continental Shelf Project, Maitama, Abuja, Nigeria

© The Author(s), under exclusive license to Springer Nature Switzerland AG 2025
S. Peppoloni and G. Di Capua, *Geoethics and Geosciences Serving Society*,
SpringerBriefs in Geoethics, https://doi.org/10.1007/978-3-032-03754-1_9

1   Introduction

Geoscience is fundamental to understanding Earth's systems and informing policy responses to global Earth and environmental challenges. It supports decisions on resource management, climate action, and risk reduction (Gill, 2017) and provides critical data that shapes international negotiations and policymaking (Gill, 2017). Diplomacy, on the other hand, facilitates international cooperation and negotiation, enabling nations to achieve shared objectives. Geoscience diplomacy has emerged at the intersection of geoscience and international relations (Hill, 2025; Raji & Stewart, 2024; Sztein, 2016). It builds on earlier science diplomacy frameworks (Plag, 2014; Gluckman et al., 2017; Kontar et al., 2018) and addresses issues in geology, oceanography and environmental changes that transcend national boundaries. Geoscience diplomacy can be categorised into three dimensions:

1. **Diplomacy for Geoscience**: Supporting international geoscientific collaboration and data sharing.
2. **Geoscience for Diplomacy**: Using geoscientific knowledge to advance diplomatic goals.
3. **Geoscience in Diplomacy**: Integrating geoscientific evidence into international policy.

Established under Article 76 of the United Nations Convention on the Law of the Sea (UNCLOS), the Commission on the Limits of the Continental Shelf (CLCS) provides a technical framework for coastal states to extend their territorial claims beyond 200 nautical miles (UN, 1982). The CLCS assesses geological, hydrological and geophysical evidence submitted by nations to establish the outer limits of their continental shelves, making it a key institution at the intersection of geoscience, policy and diplomacy (Suárez-de Vivero, 2013; UN, 1982).

The CLCS is composed of 21 experts in geology, geophysics and hydrography, elected by UNCLOS States Parties for 5-year terms (UN, 1982). Their primary mandate is to assess the geoscientific and technical validity of a state's submission for extending its continental shelf beyond 200 M (UN, 1999). The evaluation process relies on identifying the "foot of the continental slope" (FOS), which refers to the point on the ocean floor where the steep incline of the continental slope transitions to a gentler gradient. The UNCLOS uses "M" to denote nautical miles when referring to continental shelf boundaries. From the FOS, the continental margin's outer edge may be delineated by several fixed points determined by either one of two formulae specified in paragraphs 4 and 5 of Article 76. These are (Fig. 1):

1. **The sediment thickness formula (Gardiner Formula)**: A State may claim an extended continental shelf by proving that sedimentary rock thickness at fixed points is at least 1% of the shortest distance from that point to the foot of the continental slope.
2. **The distance formula (Hedberg Formula)**: The outer edge of the shelf can be defined by fixed points no more than 60 M (nautical miles) from the foot of the continental slope.

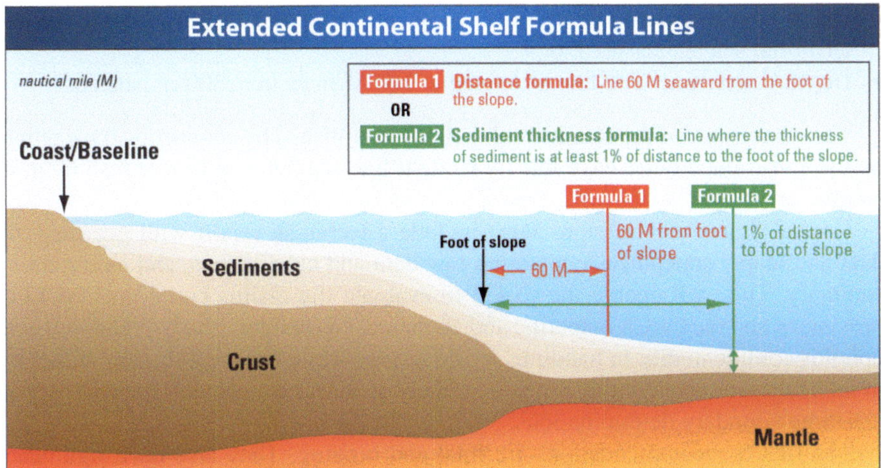

Fig. 1 Formulae for determining outer limits of the continental shelf (*Source* U.S. Department of State, Office of Ocean and Polar Affairs)

The line defining the outer edge of the continental margin is delineated by connecting those fixed points with straight lines no longer than 60 m in length. The final step in establishing the outer limits of the continental shelf involves applying the relevant constraints to the outer edge of the continental margin as specified in paragraph 5 of Article 76 (Fig. 2). There are two potential constraints:

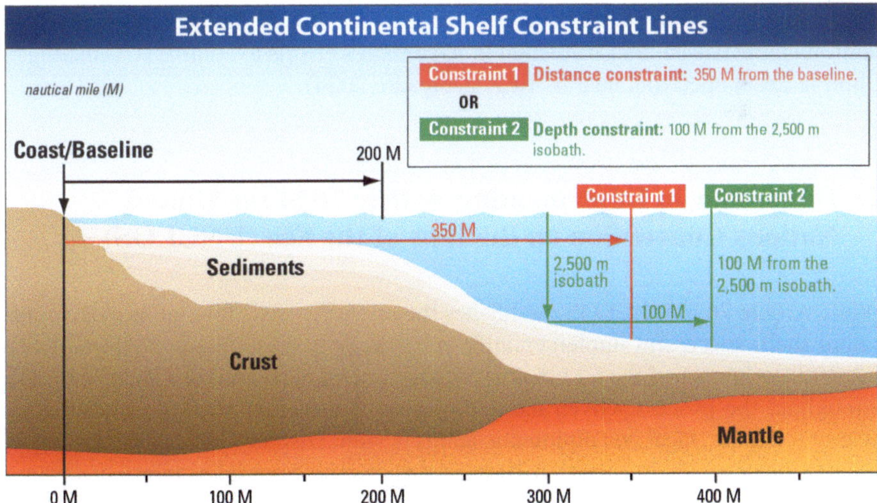

Fig. 2 Constraint lines for delineating outer limits of the continental margin (*Source* U.S. Department of State, Office of Ocean and Polar Affairs)

- **Distance constraint**: The 35 M line measured from the baselines from which the territorial sea is measured, or
- **Depth constraint**: The line is defined at 100 M from the 2500-m isobath.

The outer limits of the Continental shelf are established by connecting fixed points on the constrained outer edge line by straight line segments not more than 60 m in length.

Coastal States' submissions must include a technical report with geoscientific data and maps, appendices containing raw data and calculations, and an executive summary. This will enable the CLCS to evaluate the claims based on UNCLOS provisions and the CLCS Scientific and Technical Guidelines. The Commission also upholds confidentiality in handling sensitive data while ensuring that the summary of its final recommendations is made publicly available to maintain transparency in maritime boundary determinations.

The CLCS recommendations are final and binding. They also have significant diplomatic consequences, influencing maritime boundary negotiations, economic strategies, and geopolitical stability. Coastal states often engage in extensive diplomatic efforts to align their claims with neighbouring nations, avoiding conflicts and ensuring compliance with international law (Plag, 2014).

Beyond disputes between neighbouring coastal states, geoscience diplomacy also plays a critical role in addressing maritime conflicts involving distant nations. An example is an escalation of illegal, unreported and unregulated (IUU) fishing activities, which have emerged as one of the most pressing maritime security threats in recent years, such as China's encroachment into Argentina's exclusive economic zone (EEZ), illustrating the broader implications of maritime governance (FAO, 2020; Shi, 2023). In 2016, the CLCS approved Argentina's submission to extend its continental shelf from the standard 200 nautical miles to 350 nautical miles; this legal reinforcement has been instrumental in Argentina's efforts to combat IUU fishing by Chinese fleets operating near its EEZ (Delgado, 2024).

2 Diplomacy in Implementing Article 76 of the United Nations Convention on the Law of the Sea (UNCLOS)

While Article 76 of UNCLOS establishes the scientific basis for claims to the continental shelf, successful implementation relies heavily on diplomatic efforts. States submitting claims and the CLCS must engage in technical and diplomatic activities throughout the process. Given the complexity of these submissions, coordination among scientific, legal and diplomatic teams is vital. Disputes can emerge during the submission or evaluation phase, particularly when maritime claims intersect, or states have differing interpretations of geoscientific data. To navigate these challenges, coastal states are encouraged to consult and negotiate with neighbouring nations to ensure that proposed boundaries remain non-conflicting. Geoscience diplomacy is crucial in such scenarios as it fosters diplomatic dialogue, promotes collaborative geoscience data collection and supports cooperative solutions. Rather than escalating tensions, many states have submitted joint submissions, enabling them to assert overlapping claims while deferring the precise delimitation of boundaries (Jakobsson et al., 2012; UN, 2001). Examples of successful joint submissions include:

- The Republic of Seychelles and the Republic of Mauritius jointly secured the delineation and jurisdiction of a 396,000 km^2 extended continental shelf area in the Mascarene region (UN, 2011).
- The Federated States of Micronesia, Papua New Guinea, and the Solomon Islands jointly established the outer limits of their continental shelf in the Ontong Java Plateau region (UN, 2009).
- France, Ireland, Spain and the United Kingdom of Great Britain and Northern Ireland jointly secured the delineation of their extended continental shelf in the Celtic Sea and Bay of Biscay region (UN, 2009).
- Cabo Verde, the Gambia, Guinea, Guinea-Bissau, Mauritania, Senegal and Sierra Leone jointly secured the extension of their continental shelf in the Atlantic Ocean off the coast of West Africa (UN, 2014).

These cases show how geoscience-driven diplomacy fosters international collaboration, enabling states to work across geographical boundaries. It supports cooperative management and governance of shared marine resources, ensuring continental shelf extensions promote sustainability, legal clarity and geopolitical stability.

Successful boundary agreements often rely on bilateral and regional diplomatic agreements. The Norway–Russia Treaty, signed in 2010, resolved decades-long overlapping maritime claims in the Barents Sea and Arctic Ocean through bilateral negotiations. The agreement established a 1,750-km maritime boundary, defining the outer limits of each country's continental shelf. This treaty concluded a protracted territorial dispute and set a precedent for peaceful, science-informed resolution of maritime boundaries in the Arctic (Neumann, 2010). Regional organisations such as the Arctic Council and the Scientific Committee on Antarctic Research (SCAR) are crucial in promoting cooperation among coastal states in polar regions.[1,2] The CLCS facilitates international cooperation by bringing coastal states, scientific experts and international institutions together to address maritime boundary claims. While the CLCS issues final and binding recommendations on the technical aspects of continental shelf limits, it does not resolve political disputes or provide advisory opinions between states.

The CLCS Rules of Procedure require a high level of diplomacy from both coastal states and the Commission to ensure effective implementation. As part of the process, the CLCS appoints a Sub-commission to conduct the initial technical and scientific review of each submission. It reviews data, consults with the submitting state, and prepares draft recommendations for consideration by the full 21

---

[1] Scientific cooperation in the Arctic: Policy and sustainability efforts, Arctic Council. Retrieved from: https://www.arctic-council.org (accessed on 15 February 2025).

[2] Scientific Committee on Antarctic Research (SCAR) website. Geoscience Group. Retrieved from: https://scar.org/science/geo/geoscience/ (accessed 25 February 2025).

members of the Commission. The submission process also provides a diplomatic platform for coastal states to address and resolve issues related to their claims. This includes exchanging documents, holding meetings, and negotiating with neighbouring states to avoid overlapping or conflicting maritime boundaries. Such coordination often demands sustained, high-level diplomatic engagement. In addition to boundary issues, broader environmental and geopolitical challenges, such as marine pollution and the effects of climate change on the continental shelf, also require international cooperation. Managing resources within the extended continental shelf, including hydrocarbons, solid minerals and sedentary species, depends on diplomatic agreements that protect states' sovereign rights while fostering regional stability and sustainable development.

2.1 The Broader Impact of Geoscience Diplomacy

Geoscience diplomacy is crucial for tackling global challenges by fostering international collaboration and informed decision-making. Scientific insights into climate variability, ocean currents and greenhouse gas emissions significantly shape international climate policies and negotiations[3] (IPCC, 2021). Many vital natural resources extend across national boundaries, including groundwater aquifers, mineral reserves, and fossil fuel deposits. Geoscientific cooperation is essential to effectively manage these resources, facilitate joint research, avert conflicts and enhance diplomatic relations,[4,5,6] Transboundary natural hazards like earthquakes, volcanic eruptions, tsunamis and extreme weather require coordinated international responses. Collaborative geoscience initiatives, including shared early warning systems and hazard monitoring networks, are crucial for minimising disaster risks and safeguarding lives. The United Nations Office for Disaster Risk Reduction (UNDRR) offers global frameworks that facilitate this cooperation.[7]

Geoscience diplomatic engagement is essential for facilitating the exchange of seismic, hydrological and meteorological data across borders. This is particularly

---

[3] United Nations Framework Convention on Climate Change (2015). The Paris Agreement. Retrieved from: https://unfccc.int (accessed 18 February 2025).

[4] European Environment Agency website. Cross-border cooperation on renewable energy: https://www.eea.europa.eu/publications/cross-border-cooperation-on-renewable-energy (accessed on 8 January 2025).

[5] International Energy Agency website. Energy system—Decarbonisation Enablers—International Collaboration: https://www.iea.org/energy-system/decarbonisation-enablers/international-collaboration (accessed 10 June 2025).

[6] International Groundwater Resources Assessment Centre website. Transboundary aquifers and international cooperation: https://un-igrac.org/our-work/research-themes/transboundary-aquifers/ (accessed on 20 February 2025).

[7] United Nations Office for Disaster Risk Reduction (UNDRR) & World Meteorological Organization (WMO) website. Global status of multi-hazard early warning systems: Target G. Retrieved from: https://www.undrr.org/publication/global-status-multi-hazard-early-warning-systems-2022 (accessed 22 February 2025).

important in polar regions, where climate change has significant impacts. International collaboration through frameworks such as the Antarctic Treaty System (ATS),[8] the Arctic Council and the Scientific Committee on Antarctic Research (SCAR) ensures ongoing multinational research, environmental protection and geopolitical stability in these sensitive areas. Despite ongoing geopolitical tensions, geoscience diplomacy proves its importance by fostering trust, enabling collaborative solutions and promoting peace and stability through shared geoscientific knowledge in many regions.

2.2 Geoscience Meets Diplomacy: The Case of Nigeria's Extended Continental Shelf Submission (2000–2024)

Nigeria ratified UNCLOS in 1986, granting the country the right to extend its continental shelf if sufficient scientific evidence can be provided to demonstrate that its continental shelf extends beyond 200 nautical miles. Recognising the potential economic benefits, including oil, gas and solid mineral resources, the Nigerian government embarked on extensive collections and analyses of geological, geophysical and bathymetric data used to identify the foot of the continental slope and apply the sediment thickness formula to delineate the outer edge of the continental margin. Following years of evaluation, the CLCS issued its final recommendations in 2024, granting Nigeria an additional 16,300 km^2 of maritime area territory. In 2024, Nigeria celebrated the successful extension of its continental shelf by 16,300 km^2 beyond its 200 M limit, a milestone that reflected years of rigorous geoscientific research and sustained diplomatic engagement. This landmark decision by the CLCS is an excellent example of geoscience diplomacy in action. This newly recognised area is expected to generate substantial economic benefits, particularly in oil, gas, solid minerals and sedentary marine organisms that remain fixed to the seabed, such as clams, mussels and corals.

The success of Nigeria's submission was based on some key lessons:

- **The importance of accurate geoscience data**: Nigeria's Extended Continental Shelf (ECS) project emphasised the need for high-quality geological, geophysical and hydrographic data. Advanced technologies, including seismic, gravity, magnetic and bathymetric surveys, played a crucial role in supporting its claim.
- **Specialised expertise**: The involvement of renowned expert geologists, geophysicists, hydrographers and GIS specialists was essential in collecting and interpreting the complex geological data used for the ECS claim.
- **Diplomatic coordination**: Nigeria engaged in bilateral and regional negotiations with neighbouring states including Ghana, Togo, Benin, Côte d'Ivoire, Cameroon and São Tomé and Príncipe to address potential conflicts with these countries and

---

[8] Antarctic Treaty Secretariat: The Antarctic Treaty and related agreements, retrieved from: https://www.ats.aq (accessed 12 January 2025).

prevent future disputes that could hinder the CLCS from examining any of these States' submissions as prescribed in Paragraph 5(a) of Annex I. If invoked by one State against another State's submission, these annexes could obstruct the consideration of the submission by the CLCS.

- **Strategic planning and Governance**: Financial planning, stakeholder engagement, incorporating government agencies and local communities and project management ensured the submission met CLCS standards and gained international recognition. Effective knowledge management was critical for preserving ECS project data, information, and expertise to a high standard of confidentiality.

While implementing Article 76 of UNCLOS is primarily a scientific and technical process, diplomatic engagement remains essential in resolving maritime disputes and extending maritime claims. Nigeria's successful Extended Continental Shelf Project highlights the country's commitment to effective, transparent maritime resource governance by international law. It serves as a model for other nations pursuing similar claims.

3   Conclusion

Geoscience diplomacy is essential for addressing global challenges by connecting geoscientific knowledge with international policymaking. It facilitates cooperation, enhances decision-making and ensures geoscientific evidence supports negotiations on various issues, including maritime boundaries and environmental sustainability. One notable example is Nigeria's Extended Continental Shelf submission (2000–2024), demonstrating how geoscience diplomacy underpins evidence-based international policy and decision-making while promoting collaboration in geoscience, economic development and regional stability. In addition to maritime boundary negotiations, geoscience diplomacy is crucial for managing transboundary natural resources, reducing disaster risks and adapting to climate change. Strengthening international agreements on groundwater, offshore minerals and hydrocarbon deposits is important for ensuring sustainable resource use and minimising regional conflicts. Expanding seismic monitoring and coordinating hazard response efforts can significantly improve disaster preparedness and resilience.

As of January 9, 2025, the CLCS had received 95 submissions from coastal states and 11 revised submissions, and had issued 27 recommendations (Fig. 3). These achievements have required extensive geoscientific research and complex diplomatic negotiations. As nations face global challenges, geoscience diplomacy will remain a foundation for sustainable and collaborative international policymaking. To maximise the benefit of geoscience diplomacy, policymakers, geoscientists, multilateral organisations and diplomats should prioritise integrating geoscientific expertise into global decision-making, strengthening regional and multilateral collaboration, expanding capacity-building programmes in geoscience and diplomacy and

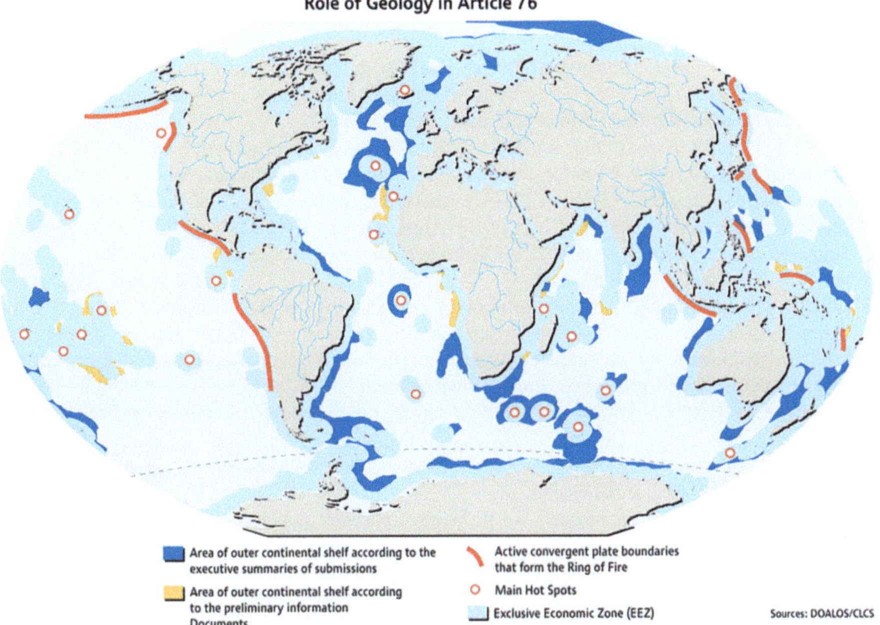

Fig. 3 World map of extended continental shelf (ECS) Areas (*Source* US Extended Continental Shelf Project Office, 2023[9])

promoting open geoscience data-sharing initiatives to foster peaceful collaboration between nations.

References

Delgado, J. (2024). Argentina Launches Operation Against Chinese IUU Fishing. Diálogo Américas. Retrieved May 20, 2025, from: https://dialogo-americas.com/articles/argentina-lau nches-operation-against-chinese-iuu-fishing

FAO. (2020). *The state of world fisheries and aquaculture 2020: Sustainability in action.* Food and Agriculture Organisation. https://doi.org/10.4060/ca9229en

Gast, A. P. (2021). A foreign policy for universities: Science diplomacy as part of a strategy. In *Science & Diplomacy. American Association for the Advancement of Science (AAAS).* Retrieved February 18, 2025, from https://www.sciencediplomacy.org

Gill, J. C. (2017). Geology and the sustainable development goals. *Episodes, 40*(1), 70–76. https://doi.org/10.18814/epiiugs/2017/v40i1/017010

Gluckman, P. D., Turekian, V., Grimes, R. W., & Kishi, T. (2017). Science diplomacy: A pragmatic perspective from the inside. *Science & Diplomacy, 6*(4). Retrieved June

---

[9] U.S. Department of State (2023). World Map of Extended Continental Shelf Areas. Retrieved from: https://www.state.gov/world-map-of-extended-continental-shelf-areas/ (accessed 27 February 2025).

10, 2025 from https://www.sciencediplomacy.org/sites/default/files/pragmatic_perspective_sci ence_advice_dec2017_1.pdf

Hill, C. (2025). Geopolicy: Science diplomacy in a new geopolitical order. *European Geosciences Union (EGU) Blog*. Retrieved February 18, 2025 from. https://blogs.egu.eu/geolog/2025/02/27/ geopolicy-science-diplomacy-in-a-new-geopolitical-order/

IPCC. (2021). Sixth Assessment Report (AR6). *Intergovernmental panel on climate change*. Retrieved June 10, 2025, from https://www.ipcc.ch/assessment-report/ar6/

Jakobsson, M., Mayer, L. A., Coakley, B., Dowdeswell, J. A., Forbes, S.,... & Weatherall, P. (2012). The International Bathymetric Chart of the Arctic Ocean (IBCAO) Version 3.0. *Geophysical Research Letters, 39*(12), Article L12609. https://doi.org/10.1029/2012GL052219

Kontar, Y. Y., Beer, T., Berkman, P. A., Eichelberger, J. C., Ismail-Zadeh, A., et al. (2018). Disaster-related science diplomacy: Advancing global resilience through international scientific collaborations. *Science & Diplomacy, 7*(2). Retrieved February 8, 2025, from https://www.sciencediplomacy.org/article/2018/disaster-related-science-diplom acy-advancing-global-resilience-through-international

Neumann, T. (2010). Norway and Russia agree on maritime boundary in the Barents Sea and the Arctic Ocean. *ASIL Insights, 14*(34). Retrieved February 15, 2025, from https://www.asil.org/insights/volume/14/issue/34/norway-and-russia-agree-maritime-bou ndary-barents-sea-and-arctic-ocean

Plag, H. P. (2014). Extreme geohazards—A growing threat for a globally interconnected civilization. *Natural Hazards, 72*, 1275–1277. https://doi.org/10.1007/s11069-014-1223-3

Raji, M., & Stewart, I. (2024). Geoscience diplomacy. *Geoscientist Online*. Retrieved January 10, 2025, from https://geoscientist.online/sections/unearthed/geoscience-diplomacy/

Shi, X. (2023). International advocacy and China's distant water fisheries policies. *Marine Policy, 152*. https://doi.org/10.1016/j.marpol.2023.105635

Suárez-de Vivero, J. L. (2013). The extended continental shelf: A geographical perspective of the implementation of Article 76 of UNCLOS. *Ocean & Coastal Management, 73*, 113–126. https:// doi.org/10.1016/j.ocecoaman.2012.10.021

Sztein, A. E. (2016). Science diplomacy in the geosciences. In J. Drake, Y. Kontar, J. Eichelberger, T. Rupp, & K. Taylor (Eds.), *Communicating climate-change and natural hazard risk and cultivating resilience*. Advances in natural and technological hazards research (Vol. 45, pp. 279– 294). Springer. https://doi.org/10.1007/978-3-319-20161-0_18

UN. (1982). *United Nations convention on the law of the sea*. United Nations. Retrieved February 22, 2025, from https://www.un.org/depts/los/convention_agreements/texts/unclos/unclos_e.pdf

UN. (1999). *Scientific and technical guidelines of the commission on the limits of the continental shelf*. United Nations. Retrieved February 22, 2025, from https://www.un.org/depts/los/clcs_ new/commission_guidelines.htm

UN. (2006). *Joint submission to the commission on the limits of the continental shelf in the area of the Celtic Sea and the Bay of Biscay*. United Nations. Retrieved February 22, 2025, from https:// www.un.org/depts/los/clcs_new/submissions_files/submission_frgbires.htm

UN. (2001). *Decision regarding the date of commencement of the ten-year period for making submissions to the Commission on the Limits of the Continental Shelf set out in article 4 of Annex II to the United Nations Convention on the Law of the Sea*, SPLOS/72, 29 May 2001. Digital Library, Available at https://digitallibrary.un.org/record/441543?v=pdf

UN. (2009). *Joint submission to the commission on the limits of the continental shelf concerning the Ontong Java Plateau*. United Nations. Retrieved February 22, 2025, from: https://www.un. org/depts/los/clcs_new/submissions_files/submission_fmpgsb_32_2009.htm

UN. (2011). *Commission on the limits of the continental shelf (CLCS) outer limits of the continental shelf beyond 200 nautical miles from the baselines: Submissions to the commission: Joint submission by the Republic of Mauritius and the Republic of Seychelles* (Adopted 30 March 2011). Available at https://www.un.org/depts/los/clcs_new/submissions_files/submission_musc.htm

UN. (2014). *Joint submission to the commission on the limits of the continental shelf with respect to areas in the Atlantic Ocean adjacent to the coast of West Africa.* United Nations. Retrieved February 22, 2025, from https://www.un.org/depts/los/clcs_new/submissions_files/submission_wa7_75_2014.htm